Prepared by the Special Publications Division
National Geographic Society, Washington, D. C.

America's Majestic Canyons

AMERICA'S MAJESTIC CANYONS

Published by
THE NATIONAL GEOGRAPHIC SOCIETY
ROBERT E. DOYLE, *President*
MELVIN M. PAYNE, *Chairman of the Board*
GILBERT M. GROSVENOR, *Editor*
MELVILLE BELL GROSVENOR, *Editor Emeritus*

Prepared by
THE SPECIAL PUBLICATIONS DIVISION
ROBERT L. BREEDEN, *Editor*
DONALD J. CRUMP, *Associate Editor*
PHILIP B. SILCOTT, *Senior Editor*
RON FISHER, *Managing Editor*
TONI EUGENE, LOUISA MAGZANIAN, *Senior*
    Researchers; BROOKE J. KANE,
    DEBORAH J. RYAN, *Researchers*
Illustrations and Design
THOMAS B. POWELL, III, *Picture Editor*
JODY BOLT, *Art Director*
SUEZ B. KEHL, *Assistant Art Director*
TURNER HOUSTON, CINDA ROSE, *Design Assistants*
JOHN BLAUSTEIN, STEPHEN J. KRASEMANN,
    DAVID MUENCH, BO RADER, *Contributing*
    Photographers
RALPH GRAY, STEPHEN J. HUBBARD,
    H. ROBERT MORRISON, MICHAEL W. ROBBINS,
    GEORGE JOSEPH TANBER,
    JENNIFER C. URQUHART, SUZANNE VENINO,
    EDWARD O. WELLES, JR., *Picture Legends*
JOHN D. GARST, JR., CHARLES W. BERRY,
    MARGARET DEANE GRAY, MARK H. SEIDLER,
    TIBOR G. TOTH, CATHY WELLS,
    ALFRED L. ZEBARTH, *Map Research,*
    Design, and Production
Production and Printing
ROBERT W. MESSER, *Production Manager*
GEORGE V. WHITE, *Assistant Production Manager*
RAJA D. MURSHED, JUNE L. GRAHAM,
    CHRISTINE A. ROBERTS, DAVID V. SHOWERS,
    Production Assistants
DEBRA A. ANTONINI, BARBARA BRICKS,
    BETSY BURKHOLDER, JANE H. BUXTON,
    KAY DASCALAKIS, ROSAMUND GARNER,
    SUZANNE J. JACOBSON, CLEO PETROFF,
    KATHERYN M. SLOCUM, SUZANNE VENINO,
    Staff Assistants
ELIZABETH MEYENDORFF, *Index*

*Cradling a newborn calf, Paul Johnson
strides across his farm, acreage that nestles
among the hills of the Upper Iowa River
gorge. Pages 2-3: The Rio Grande flows from
the yawning mouth of Santa Elena Canyon
on the border of Texas and Mexico. Page 1:
Yosemite Falls drops 2,425 feet to a glacier-
carved valley floor. Bookbinding: Hay-
scented fern, found in Pennsylvania's Pine
Creek Gorge, unfurls a tightly curled frond.*

MATT BRADLEY. PAGES 2-3: TOR EIGELAND.
PAGE 1: N.G.S. PHOTOGRAPHER JOSEPH H. BAILEY

Wrinkled and scarred, the face
of North America bears the
marks of more than a billion years of
geological activity. Uplifting,
erosion, glaciation, folding and
faulting, heating and cooling, river
carving, and weathering by wind
and rain have all collaborated to
create a continent seamed with
cracks and fissures. Geologists
list more than a thousand
canyons in North America and
Hawaii, ranging from gentle,
wooded slopes of the Midwest
and East, to astonishing,
rock-walled chasms of the West.
For inclusion in this book,
geologists and National Geographic
staff members selected a
representative sampling of
canyons, gorges, and valleys,
basing their choices on geological
interest, scenery, uniqueness,
and geographic distribution. In
each region, one canyon (gold
box) receives major attention;
those remaining combine in
smaller but dazzling portfolios
of America's majestic canyons.

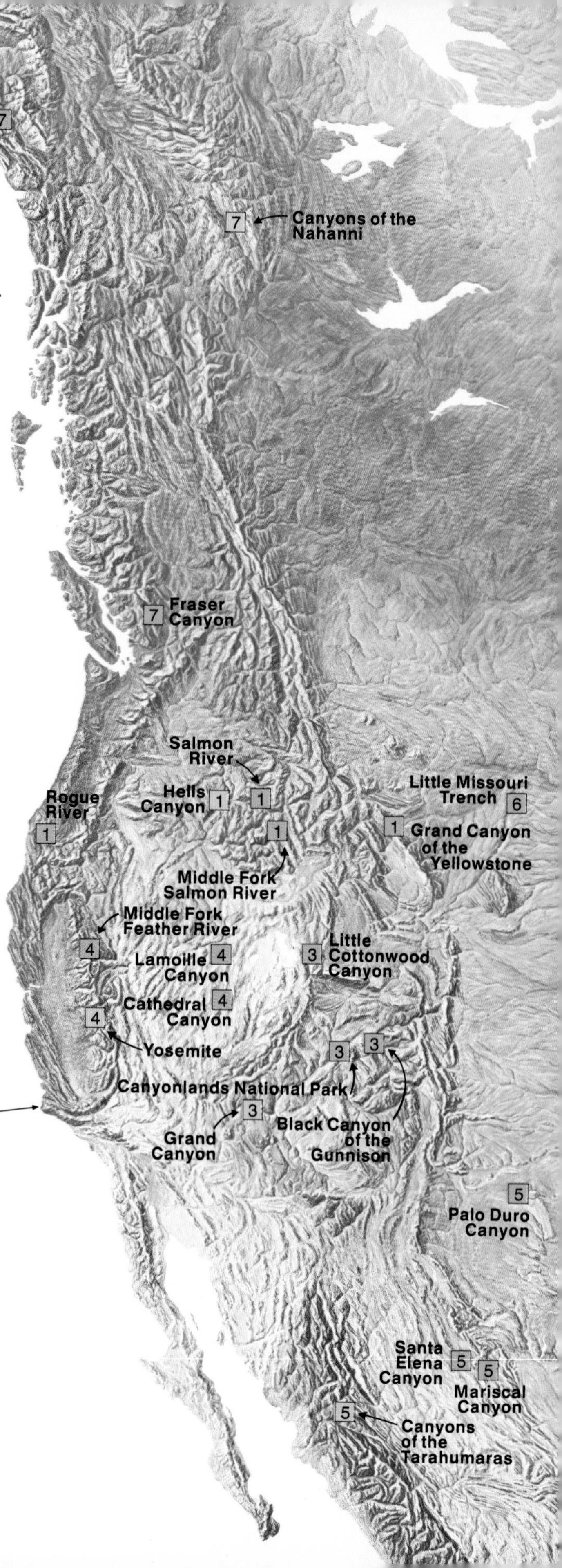

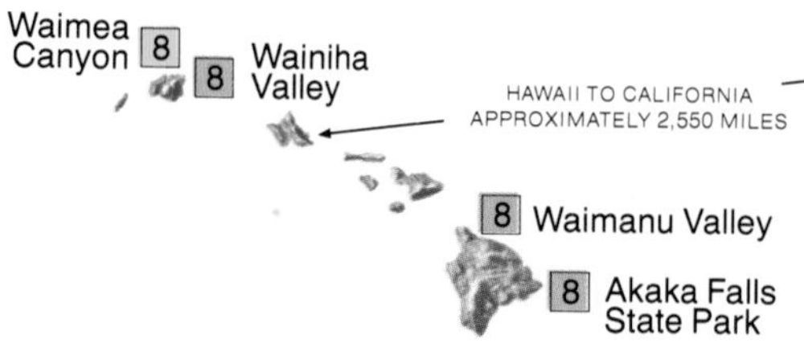

Wood Canyon 7

Canyons of the Nahanni 7

Fraser Canyon 7

Salmon River

Hells Canyon 1   1

Rogue River 1

Little Missouri Trench 6

Grand Canyon of the Yellowstone 1

Middle Fork Salmon River 1

Middle Fork Feather River 4

Lamoille Canyon 4

Little Cottonwood Canyon 3

Cathedral Canyon 4

Yosemite 4

Canyonlands National Park 3   3

Grand Canyon 3

Black Canyon of the Gunnison

Waimea Canyon 8

Wainiha Valley 8

HAWAII TO CALIFORNIA
APPROXIMATELY 2,550 MILES

Waimanu Valley 8

Akaka Falls State Park 8

Palo Duro Canyon 5

Santa Elena Canyon 5   5

Mariscal Canyon

Canyons of the Tarahumaras 5

7 Agawa Canyon

6 Dalles of the
St. Croix

Gorge of the 6
Upper Iowa

2 Niagara Gorge

2
Pine Creek
Gorge

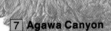
New River Gorge 2

6 Jacks Fork River

2 Linville
Gorge

The Northwest:

Bright with spring grasses, rocky ledges along the Snake River rise to only a fraction

Hells Canyon

Photographed by Phil Schofield

of the staggering 7,900-foot height of Hells Canyon—the deepest gorge in North America.

Sparks streak the night as author Will Gray (second from left) and companions gather around a campfire. During a seven-day float trip down the Snake River, the current set the pace as their rafts drifted through Hells Canyon. Guided by outfitters Jerry Hughes and Carole Finley, they coursed playful riffles and raging torrents of white water. On frequent side trips, the party explored the countryside and climbed rocky cliffs for views of the canyon. At night, they camped along the riverside, cooking hearty meals and later joining in song. Embarking downstream of Hells Canyon Dam, they followed the Snake along the Oregon-Idaho border to the mouth of the Grande Ronde River in Washington. Three dams now impound river waters of the canyon; responding to protests by environmentalists against more, Congress established the Hells Canyon National Recreation Area in 1975. After a long day, Jerry cooks the traditional last meal of the trip: forty-mile stew. First browning beef and onions in Worcestershire sauce, he adds carrots, potatoes, peppers, corn—improvising with leftovers—then simmers the stew for about an hour. He tops it off with a layer of cheese.

WILLIAM R. GRAY, NATIONAL GEOGRAPHIC STAFF (THIS PAGE)

Hot biscuits will soak up every last bit of meaty gravy. After preheating a Dutch oven over smoldering coals, Carole places the biscuits in an oiled pot, covers it, then heaps coals on the lid. In half an hour the biscuits bake fluffy and brown. "A wilderness gourmet's delight," says Will.

Bellowing loudly, sheepherder Bill Sanderson corrals spring
lambs for docking and branding at a ranch near Kirkwood Creek
in Idaho. If not docked, the sheep would soil their wool by wagging
dirty tails. At left, a hired hand separates a young ram from
the flock for neutering. Rancher Andy Dahlquist (in striped T-shirt)
explains: "Male sheep not intended for breeding are castrated.
Then we paint a 'Bar 20' brand on all the sheep." At left,
the animals fatten on rich summer grass. For thousands of years,
Indians hunted and fished in Hells Canyon. Later, prospectors
and miners attempted to earn a living here. Some liked the mild
winter climate and stayed to homestead and raise livestock.
Today, ranching represents one of the few commercial enterprises
allowed in the largely federally-owned canyon. Ranchers must
obtain permits from the U. S. Forest Service.

Wide-open door invites only the elements and inquisitive hikers into an abandoned homestead cabin near Bernard Creek. Inside, a wood-burning stove rusts in the kitchen (right). About ten such cabins remain in Hells Canyon—remnants of an era when ranchers settled here. Early explorers and trappers visited the rugged canyon, and homesteaders overcame the hardships, tending their livestock and rearing families in the wilderness. Below, afternoon sunlight fills the Barton cabin near Battle Creek. Ace Barton, 54, who now lives in nearby Riggins, Idaho, recalls: "It was a hard life, looking back on it. But at the time we didn't know any better, so it seemed pretty good." His grandfather homesteaded in Hells Canyon in 1902, and the Barton family raised cattle here until selling out to sheep ranchers in 1952. "We had to be pretty much self-sufficient," says Ace, "since we were so far from town. If something broke, you fixed it. You had to learn to be blacksmith or mechanic or whatever. You couldn't just run to the hardware store. Sometimes we didn't see any other people for months, but we had a feeling of closeness and community that's hard to find nowadays. I sure miss a lot of things from those days—the solitude, the freedom, the feeling of self-reliance."

Past bluffs wrinkled by time, a raft drifts down the Snake River; the current carries boaters beside sweeping green tributary valleys and sheer walls of volcanic rock. Millions of years ago, volcanoes in Idaho erupted and poured molten lava over the land. Since then, alternating forces of volcanism, uplift, and erosion have gradually created the canyon. From its arid depths to its alpine peaks, Hells Canyon supports a wide variety of wildlife. Above, two mule deer freeze at the sound of rustling grass. A land snail searches for food, and a bull snake, often mistaken for a diamondback because of its pattern, skims across sunlit water.

Preceding pages: Cold splash of white water soaks Carole Finley as she runs Granite Rapids. Through rushing backwaters, whirlpools, and waves cresting up to ten feet, she skillfully maneuvers the raft. "We always scout the major rapids before running them," says Carole. "The Snake has powerful high water, and the level of the river constantly fluctuates because of the dams, snowmelt, or heavy rains. So your approach to a rapid can differ with each run." Landing upstream of the rapid, the guides check it for water level, obstacles, and various types of waves. "We read the river carefully," Carole continues, "and then choose a path where the current will do most of the work."

Dark clouds lower in the distance, but sunshine lingers on hikers tramping through tall grass above Cook Creek. The steep slopes of the side canyon required frequent switchbacking and tested the hikers' muscles. A stately grove of ponderosa pine (right) mantles a slope and scents the air with a pungent perfume. Below, Jerry Hughes bends low for a drink from Battle Creek.

Wild flowers add a bold stroke of color to
Hells Canyon. Beneath snowcapped peaks,
sunflowers brighten a mountain meadow (right).
The three-lobed deer horn (below), named
Clarkia for its discoverer, the co-captain
of the Lewis and Clark party, turns slopes shocking pink
in early summer. White sego lilies sustained
Indians and pioneers; their small bulbs taste
like potatoes. Daisylike asters grow in clusters
of lavender, pink, blue, or white throughout the West.

Parasitic broomrape
(top) feeds on the roots
of other plants, and lacks
chlorophyll to green its
yellow stems. Dew jewels the
elegant blossoms of a scarlet
gilia; it may startle hikers
with its skunklike odor.

By
William R. Gray

Precipitous walls, brushed green with spring grasses, plunged toward a ribbon of water sparkling nearly a mile below me. Pewter clouds banked the snow-covered tops of the walls, which reared overhead. Only from the vantage of a single-engine plane could I appreciate the full, dramatic scale of Hells Canyon—and even then only by craning my neck.

At its deepest point it drops 7,900 feet, ever-narrowing, from the craggy heights of the Seven Devils Mountains to the churning rapids and eddies of the Snake River. This magnificent canyon, which ranges from an alpine rim to an arid floor, is the deepest gorge in all of North America; it eclipses the Grand Canyon by 2,000 feet.

Despite its staggering size and breathtaking beauty, Hells Canyon is but one of the truly great gorges in the northwestern tier of states. I explored the rocky chasms of the dancing Rogue River in Oregon, the pine-flanked canyons of the Salmon and the Middle Fork Salmon rivers in Idaho, the vivid gash carved by the Yellowstone River in Wyoming.

Throughout the Northwest I found canyons of majesty, canyons of beauty—but to me none of the others combined the elements of history, geology, wilderness, wildlife, and scenic wonder in the dramatic style of Hells Canyon. My reactions mirrored those of Capt. Benjamin L. E. Bonneville, one of the West's early explorers. Nearly a century and a half ago he ventured into the Hells Canyon area with a group of comrades. In 1837 Washington Irving published an account of the trip in *The Adventures of Captain Bonneville, U.S.A.:* "The grandeur and originality of the views, presented on every side, beggar both the pencil and the pen. Nothing we had ever gazed upon in any other region could for a moment compare in wild majesty and impressive sternness, with the series of scenes which here at every turn astonished our senses, and filled us with awe and delight."

Also with a group of comrades, I ventured into Hells Canyon, but instead of bumping along on horseback we drifted down the river in rafts. I joined outfitters Jerry Hughes and Carole Finley, husband-and-wife owners of Hughes River Expeditions, Inc., of Cambridge, Idaho, for a week-long float through the canyon. Jerry, a robust 30-year-old with tousled hair and a quick smile, has been guiding river runners throughout the West for 13 years. Although Jerry has a degree in law, he and Carole—a delicate woman with flowing, strawberry-blonde hair—decided to devote their careers to the rivers they love, including the Snake.

Joining Jerry, Carole, and me were boatman Gil Hagan and several members of the staff of this book: map researcher (and my wife) Margie Deane, researcher Toni Eugene, picture editor Tom Powell and his lawyer-wife Barbara, and Idaho-based free-lance photographer Phil Schofield. On a cloudless June morning we gathered just below Hells Canyon Dam, a massive monolith of concrete completed in 1968, to begin our trip. Three such dams block the waters of the Snake within 40 miles of our put-in point. Proposals to build more dams downriver prompted the designation of the canyon as a national recreation area. As we loaded our rafts, Jerry glanced wistfully up at the dam and said, "Five of the seven biggest and most challenging rapids in Hells Canyon are buried beneath the lake

Clouds part to reveal the austere majesty of Hells Canyon. This gaping abyss measures nine miles across at its widest point, from the summit of He Devil Mountain in Idaho to the western rim in Oregon. Plummeting nearly a mile and a half, Hells Canyon could hold six Empire State Buildings stacked one atop the other—with room to spare.

up there. Like a lot of people, I never had a chance to run them. I'm just glad that the rest of the canyon is protected."

To whet our appetites for the two monstrous rapids that remain, we glided through some riffles that turned the green waters of the Snake to foaming white. As we bobbed along, I leaned back and surveyed Hells Canyon. Terraced bluffs of gray, brown, and black rock soared skyward in steps of several hundred feet each. As towering as they seemed, I knew the bluffs were only a fraction of the canyon's total height, for it is measured from the river to the top of He Devil Mountain, five miles to the east. The soft greens and golds of the shrubs and grasses mantling the rock contrasted sharply with the cobalt blue of the sky and the bright, nearly white-hot burst of the sun.

The bluffs reminded me of an old legend that explained the formation of Hells Canyon and the Seven Devils Mountains. Long ago, seven powerful giants stalked the land, marauding villages and devouring children. To defend themselves, the people sought the help of Coyote, an animal-person with mystical powers. With the assistance of Fox, Coyote had seven deep pits dug and filled with a boiling reddish-yellow liquid. Soon the giants marched near, their heads held haughtily in the air, certain that nothing could surprise them. Suddenly, they slipped into the pits and sank deep into the hot liquid. They struggled, but succeeded only in spilling the fluid over the land for many miles. Finally, Coyote came out of hiding and changed the giants into mountains. To keep any more enemies from crossing into the land, he created a deep trench in front of the mountains—Hells Canyon.

Like many legends, this one contained a nugget of truth, I found, when I talked with geologist George Williams at the University of Idaho. There were indeed once volcanoes in the Seven Devils area, and the liquid in the legend could have been lava.

"Volcanism has marked the Hells Canyon region for hundreds of millions of years," said Dr. Williams, a sprightly scientist with a salt-and-pepper goatee. "Eruptions blanketed the area with thick flows of lava. After a long, complex period of erosion, uplifting, deposition of more rock, and more erosion, huge floods of lava from the west poured across the land. Layer upon layer, basalt was piled miles thick. In fairly recent times, geologically speaking, the Snake River began cutting into the rock; at the same time, the land was being uplifted. The result is the vast erosional canyon that we see today."

A splash of cold Snake River water jolted me from my thoughts. I glanced upstream and for a moment watched our small flotilla bounce through the sunlit canyon. Carole, I noticed, worked the oars of her raft with the precision of a veteran. "Carole is really something," Jerry said. "She's about the only woman out here who will consistently run Hells Canyon. The river's so powerful and so deep that it scares people off—but not Carole. She's not as strong as I am and doesn't have the same leverage on the oars because she's not as tall. But she makes up in intelligence and finesse what she lacks in strength. She works her way through a rapid, dodging here, cutting there, while I just bull my way through."

I had a chance to judge Carole's technique as we neared Wild Sheep Rapids—usually the largest in Hells Canyon. A steadily increasing roar, the noise of the roiling water reverberating against the canyon walls, signaled our approach. Jerry waved all the boats to the bank so that he, Carole, and Gil could scout the rapid. "We always look at the big ones

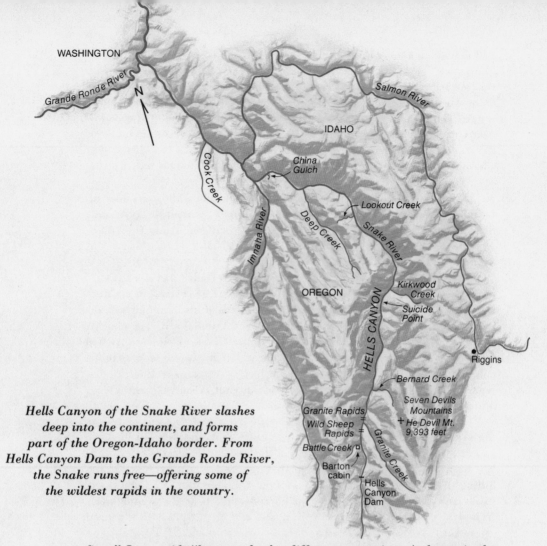

WASHINGTON

Grande Ronde River

N

IDAHO

Salmon River

Cook Creek

China
Gulch

Lookout Creek

Imnaha River

Deep Creek

Snake River

OREGON

Kirkwood
Creek

Suicide
Point

HELLS CANYON

Riggins

Bernard Creek

Seven Devils
Mountains

He Devil Mt.
9,393 feet

Granite Rapids

Wild Sheep
Rapids

Granite Creek

Battle Creek

Barton
cabin

Hells
Canyon
Dam

*Hells Canyon of the Snake River slashes
deep into the continent, and forms
part of the Oregon-Idaho border. From
Hells Canyon Dam to the Grande Ronde River,
the Snake runs free—offering some of
the wildest rapids in the country.*

first," Jerry said, "because they're different every time. A change in the level of the water, for instance, can completely alter the way you approach a rapid like Wild Sheep. If you go in blind, you're asking for trouble."

I walked downstream with them, carefully avoiding the spreading growths of poison ivy that decked the canyon. Some individual plants reached above my head, and in certain exposed areas the ivy covered hundreds of square yards.

After checking Wild Sheep and discussing the possible alternatives, the boatmen decided which routes to take. We returned to the rafts and cast off. I rode in the lead boat with Jerry, along with Margie, Toni, and Lady, Jerry's yellow Labrador retriever. She reclined regally on a pile of duffel and simply yawned at the approaching white water.

As we neared the lip of the rapid, small waves began to pluck at the bottom of the raft, then suddenly we were pitched into the middle of the torrent of splashing, crashing water. We edged down the front of a huge wave and along the perimeter of a "hole," a steep slick of sea-green water that ended in a towering, back-curling breaker. The wave stood ten feet high and would have flipped the raft if we had strayed into it. Straining against the oars, Jerry worked us sideways behind that mammoth wave and we roller-coastered through the rest of the rapid, bounding, jouncing, and getting soaked by the splash.

We pulled for the bank to bail, and I watched Carole ease her raft into the rapid. With a few flicks of her oars, she was past the dangerous hole

27

and skittering among the standing waves. After Gil successfully ran Wild Sheep, we floated a mile or two to our first campsite—Granite Creek, a roaring tributary choked with silt. A small ledge with a rock overhang made a perfect kitchen, and Jerry and Carole soon had steaks sizzling over a fire. That night, our sleeping bags spread in a flower-dotted meadow near Granite Creek, Margie and I gazed in wonderment at the brilliance of the stars. They blazed with crystalline clarity against the blue-black sky. The only distraction from the starscape was an occasional bat, darting overhead on silent wings.

The next morning I strolled through the meadow and part way up Granite Creek. Not far from this tranquil spot, I recalled, a mysterious double murder had occurred. Ace Barton, a third-generation Hells Canyon rancher and grandson of one of the victims, had told me the story at his home in Riggins, Idaho.

"Martin Hibbs, my grandfather, first came to Hells Canyon in 1897," said Ace. Wisps of smoke trailed from his pipe as he relaxed in a sitting room filled with arrowheads and other mementos of his life in Hells Canyon. "He discovered copper at the mouth of the Imnaha River, sold his claim, and bought some land around Granite Creek. Well, he ranched there for years and raised a family. One June day in 1934 he rode up to his cabin after spending a few days in Riggins buying supplies. He had just dismounted and taken off his spurs when he was shot in the back and killed, apparently without knowing what hit him.

"The cabin was burned down, and in the ashes another body was found—that of a miner named Joe Anderson. Nobody ever solved those murders. It was so remote out there that the law didn't seem too interested. But we in the family always thought that Joe Anderson's partner did the killing. We figured that the two miners must have gotten in a fight at the cabin and that my grandfather just happened up at the wrong time. Nobody ever knew the other miner's name, and he just plain disappeared after that." Ace shook his head and stared off into space. "It's one of those tragedies that seem to happen in places like Hells Canyon."

That night we camped at the mouth of Bernard Creek. An old cabin stood like a sentinel of the past on a bluff overlooking the Snake. Once filled with life, the cabin now sheltered only spiders, whose delicate webs filigreed open spaces along the roof. And in the surrounding meadow, weeds grew tall among pieces of rusting farm machinery. I recalled Ace Barton's observation: "It took a long time and a lot of hard work to make homesteads down there comfortable. It amazes me how quickly they've begun to revert to nature."

Nearly all of the old homesteads in Hells Canyon have been purchased over the years by the U. S. Forest Service, most before 1975 when Congress designated the canyon as part of Hells Canyon National Recreation Area. A team from the Forest Service is working to develop a comprehensive management plan for the recreation area. Jim Hulbert, one of the members of that team, joined us for part of our trip. Young and bearded, Jim told me about some of the problems confronting the group in its efforts to draw up plans for such a large and diverse area.

"First of all," he said, "two states and five counties have jurisdiction in the area. Also, three separate national forests have been administering the land for several decades. Pulling all those interests together has been difficult, but that's only part of the problem. There are several practical and emotional public issues that we have to work out.

"For instance, there's a conflict between people who float the river and those who use powerboats. We have to find ways to provide opportunities for both. Then there are controversies about logging, grazing of cattle and sheep, the amount of wilderness to set aside, and the construction of roads for canyon overlooks. The list seems endless, but we're trying to get input from everyone concerned. We hope to present a balanced approach to Congress by December 1980."

For the next couple of days we lazed through Hells Canyon, mesmerized by the ever-changing scenery. At places like Suicide Point, the canyon walls pinched together to form beetling cliffs that rose hundreds of feet above us. Elsewhere the canyon broadened into gently rising hills.

Events of those days run together in a kaleidoscope of images: Carole innocently splashing water on Jerry's boat and somehow provoking a laughing, brawling water fight; Phil drifting alongside the rafts to take water-level photographs; Lady waking everyone in the morning with sloppy kisses; me vainly trying to row a raft.

One broiling afternoon we stopped to camp near Lookout Creek. A grove of leafy hackberry trees offered a small enclave of shade, and Margie propped herself against the trunk of one to read. At dusk Phil walked past the same tree and heard a loud whirring. He called me over, and I quickly recognized the telltale sound of a rattlesnake. The snake, which had slept through the heat of the day, was wide awake now, and rattling ominously. Phil and I jumped to the top of a nearby table; we were quickly joined by several others.

Eventually Gil, dressed only in shorts and sandals, corralled the three-foot-long snake, trapping it—still fighting mad—between a shovel blade and a stick. Carefully, he carried it down to the river and threw it in. "It'll be swept through the next rapid and then swim to shore," Gil said. "I'm glad we didn't have to kill it."

Neither Margie nor Phil slept a wink that night, each continuously imagining they heard the rustle of snakes in the grass. The next morning Phil reported, "Once I was just drifting off when a gust of wind caught the corner of my tarp and blew it across my face. I thought for sure it was that rattlesnake's mate and entire family coming to get me."

The same gust of wind that disturbed Phil also indicated a change in the weather. The morning had brought a sullen gray overcast that would stay with us the rest of the trip.

The brooding sky and chill wind formed a gloomy backdrop for our lunch stop at the mouth of Deep Creek, and seemed to heighten the tragic overtones of the place. Accounts differ in details, but in 1887, thirty-one Chinese placer miners were brutally murdered near here. The Chinese had begun mining in the area in the fall of 1886, and stories soon circulated that they had 17 flasks of gold. One afternoon a gang of drifters running cattle on the Oregon rim of the canyon began firing down into the miners' camp, picking them off, one by one. Within minutes the Chinese were dead.

The cowboys tossed the bodies into the Snake and began to search for the gold. It's not known whether or not they found it. When the bodies were discovered floating past Lewiston, Idaho, an investigation was undertaken, but no one was ever convicted of the crime. And no one ever found all that gold, if it existed at all.

After lunch, I wandered up the creek. As I walked, I kicked rocks, scattered sand, and peered behind bushes, secretly hoping I might spot a moldering flask glittering with gold.

That night we camped on a long sandbar near China Gulch. The skies threatened rain, and to warm us and perhaps to cheer us, Gil built a wilderness sauna. With oars, ropes, and large plastic tarps, he fashioned a low tent just large enough for four people. After dinner, we entered the sauna with a bucket of hot cobbles from our campfire. Seated cross-legged, Gil ladled water from another bucket onto the stones. Immediately, a cloud of steam permeated the sauna. With each cup of water the steam in the room became hotter and thicker. The temperature and humidity soared, and soon I was dripping with perspiration. After several minutes we burst from the sauna, raced to the river, and dived in. I expected to be shocked by the cold, but the Snake felt pleasantly warm. I had an invigorating swim before joining my friends around the fire.

Clouds and intermittent sprinkles followed us for the next day and a half to our take-out point at the junction of the Snake and the Grande Ronde rivers. Along the way the character of the land changed as the slopes of the walls became gentler. The character of the Snake changed, too, with the addition of water from two major rivers, the Imnaha and the Salmon. The latter, one of the longest undammed rivers in the U. S., was still at flood stage. Gray with silt and heavy with logs and debris, it seemed to double the size of the Snake, and to add to it a darkly ominous cast.

Nature used a much softer stroke in southwestern Oregon, where the Rogue River winds through the rumpled Siskiyou Mountains. For four days I followed a trail down the Rogue, sometimes hiking alongside its swirling current, sometimes climbing high above it on rocky cliffs. Like author Zane Grey, I soon came to love the rhapsodic beauty of the Rogue. Grey based one of his Westerns, *Rogue River Feud*, here and even bought a cabin on Winkle Bar—a shelf of sand and gravel shaded by towering trees. In *Tales of Fresh-Water Fishing* he wrote of Winkle Bar, "I was content to walk around under the oaks and pines, to breathe the fragrance of the forest once more, to listen to the singing river, to watch the flight of wild fowl and hawks, and to gaze long at the sunset-flushed clouds above the lofty peaks in the west. . . . At all hours of the day this place was beautiful. . . ."

I found some of the same peace along the Rogue. Squirrels, deer, and birds were frequent companions; side streams chuckled with clear, sweet-tasting water; delicate irises carpeted the trail with shades ranging from deep purple to faint yellow; a bright sun warmed the days, and an icy-white moon graced the nights.

Late one afternoon the trail led me into one of the steepest sections of the Rogue River Canyon. Swirling and fuming, the river dived for half a mile between the vertical walls of Mule Creek Canyon. The green color of the Rogue seemed to deepen in comparison with the greenish-gray of the volcanic rock it had eroded. The narrow walls, no more than 50 feet apart, confined the river and brought it to a boil.

The trail wound high along the north wall of Mule Creek Canyon, then dropped to water level at Blossom Bar, site of the trickiest rapids along the Rogue. I planned to camp there that night, and after spreading my sleeping bag I boulder-hopped down to the rapids. In a torrent of white water, the Rogue plunged over and around dozens of large rocks, some the size of trucks. What would normally be a challenging run was made more difficult by the wreckage of an abandoned raft that was pinned by the current to one of the rocks in the main channel.

Several hundred yards upstream I spotted a group of women who were scouting Blossom Bar. They soon disappeared, and in a few minutes

Rogue River

two rafts and a couple of small kayaks pushed out into the current. The larger raft, with eight paddlers aboard, was the first to attempt the run. It eased into the rapids, tried to circle upstream of the wrecked boat, became snared by the current, and was swept backward over a barely submerged boulder. Two people were tossed out of the raft as it careened crazily in the rushing water. In an instant both it and the two swimmers, floating easily in their life jackets, whisked past.

Next came the two kayakers, and their light and maneuverable craft managed the rapids without mishap. Only the small raft, loaded with supplies and rowed by a single woman, remained. She drifted tentatively into the current and tried to follow the same course as the larger raft. Suddenly her craft turned on edge and both oars flew into the water. For seconds, it seemed, the raft poised at the point of turning over. Then it flopped back right side up, its passenger still aboard, and flashed by, bumping off rocks and spinning in the current.

I arose early the next morning and caught up with the women just as they were ready to embark for the day's float. Marilyn Papich, the captain of the smaller raft, invited me to ride with her for a mile or two. As we pushed off, I asked Marilyn how her group had come together. "We're just a bunch of friends who like rafting and the outdoors. This is the first trip we've made that's been all women—and we're really enjoying it. Last night we had a great time reliving our adventure at Blossom Bar. There's a lot of togetherness. We're all helping each other rather than relying on someone else to do the heavy work. I think we're going to do more of this."

After Marilyn dropped me off, I hiked to Flora Dell Creek, a sparkling tributary that cascades over a 30-foot waterfall. I lay on a shelf of rock and watched puffy clouds stray across the sky. Green dominated all other colors around me. The river was shamrock, the moss chartreuse, the leaves above me olive, and the faraway hills shades of smoky green.

Salmon River

Middle Fork Salmon River

Spring snow still thickly decked the high mountains of central Idaho when I landed at a grassy airstrip near the head of Impassable Canyon, a plunging granite gorge that frames a section of the Middle Fork Salmon River. Soon two rafts skimmed into sight, carrying Carole Finley and Jerry Hughes, Lady, and five guests. I would join them for two days on the Middle Fork, and then Jerry and I would continue down the main Salmon for another five days.

In the first half mile we swept through Haystack Rapid, a flume of white water guarded by looming boulders, and entered Impassable Canyon, named by Army Capt. Reuben Bernard. In 1879 Bernard, commanding a company of horse soldiers, penetrated the canyon to pursue the Sheepeater Indians there. The Army's stated reason for the action was that five Chinese miners who had been killed might have been victims of the Sheepeaters, a generally unaggressive band whose predecessors had lived in the Salmon River Mountains for 8,000 years. After months of arduous travel, Bernard and his men finally exhausted and brought in the Sheepeaters—a forlorn group of 51 people, only 15 of whom were warriors. The Indians were resettled on a reservation in southeastern Idaho.

At Rattlesnake Cave, a rocky cavern hollowed by the force of the river, Jerry showed me part of the legacy of the Indians. Painted on the shadowy rock walls in a dark, bloodred color were images of people and animals. Many of the pictographs had been obscured by smoke from campfires, but one, the largest, remained clear. A group of hunters, armed with spears and bows, surrounded a fiercely defiant elk. Here, I thought, is a depiction of some basic elements of the Sheepeaters' existence: a need—probably desperate at times—for food; a need to band together to find and kill that food; and an apparent need to portray the animals of their world.

I heard a shout and turned to see Jim Smith, a young physician from Boise, pointing to a grassy ridge across the river. Grazing in solitary majesty was a bull elk, its spreading antlers similar to those of the elk in the pictograph. Although the Sheepeaters are gone, I reflected, at least some of their world remains.

From a lofty bluff redolent with sage, I found an all-encompassing view of that world. Snow-topped mountains pierced a dome of sky dotted with scudding gray clouds. Rocky canyon walls, decked with grass and brush, gave way far below to the winding furrow of the Middle Fork. Ponderosa pines, their reddish, corrugated trunks rising arrow-straight, flanked the river, which was studded with moss-draped rocks. Such gentleness, I found, marked the entire Middle Fork and Salmon rivers.

Grand Canyon of the Yellowstone

If green had pervaded the Rogue River Canyon, a rainbow of pastels decorated the Grand Canyon of the Yellowstone River in Wyoming. Hues of saffron, cream, peach, tan, and cinnamon streaked the walls of this fantastic gorge, which plunges from the peaks of rolling pine-draped mountains 1,500 feet to the milky green of the Yellowstone.

My favorite view of the canyon is from the brink of the Lower Falls of the Yellowstone. Here the river plunges 309 feet over a resistant bed of lava. By leaning far out over the guardrail, I could peer straight down the falls—a hypnotic vision of rushing, frothing water. Down, down the water tumbled, turning to spray in the upward sweep of air. Mist obliterated the very bottom and billowed in undulating clouds, perpetually greening the canyon walls where it settled.

The unusual colors of the canyon, I learned, result from the same geological forces and events that have made Yellowstone National Park famous throughout the world. A fracture zone in the volcanic rock of this area permitted steam and hot water to percolate toward the surface from deep within the earth. A combination of the heat, liquid, and corrosive chemicals carried by the steam and water colored the rocks and gradually altered the texture of the lava. During many thousands of years, hard rock was softened and became so crumbly that it was easily eroded.

One cloudy July morning I hoisted my day pack and stepped off on a five-mile hike to the bottom of the canyon. The first three miles took me through dense pine forests along the lip of the gorge, and I caught sporadic glimpses of its bright walls. At one spot I saw the slender trace of Silver Cord Cascade, a dainty wisp that drops down the canyon wall.

Finally I began to descend into the canyon—1,250 feet straight down—on a steadily switchbacking trail. I passed through a sputtering hydrothermal area, crossed a creek of hot water, and soon reached the Yellowstone. At the bottom of the canyon the colors seemed more vibrant, more overpowering. I felt that I was part of a giant stained-glass window.

After exploring for a while I perched on a riverside boulder, watching a solitary osprey circling skyward from its nest on a rocky pinnacle high above me. I turned my back on the river and began the long climb out. And with each step upward I envied the osprey its soaring, effortless flight.

Thundering over Lower Falls, the Yellowstone River plunges 309 feet to the floor of the Grand Canyon of the Yellowstone. This golden chasm stretches 20 miles in northwestern Wyoming.

DAVID MUENCH

Rogue River

Autumn leaves shimmering in morning sunlight reflect the quiet beauty of the Rogue
River Canyon. Delicate flowers, abundant wildlife, and ancient cliffs cloaked
with evergreens softly harmonize as the river glides through southwestern Oregon on its
course to the Pacific Ocean. In spring, an iris (right) blooms at the river's edge, and
a praying mantis blends with its leafy surroundings. A popular recreational area, the canyon
lures thousands of vacationers each year. Sportsmen fish for steelhead trout and salmon,
hikers explore river trails, and white-water enthusiasts challenge boisterous rapids.

WILLIAM R. GRAY, N.G.S. STAFF; BUDDY MAYS (RIGHT); AND © JOHN BLAUSTEIN 1979 (TOP)

Salmon River
Middle Fork Salmon River

Kelly-green moss patterns boulders lining the banks of the Salmon River in central Idaho. Winding 425 miles, largely through national forests, the Salmon carves a trench that plunges more than a mile deep for a distance of 180 miles. Early explorers called it the "River of No Return" after futile attempts to travel its wild water. Today, the Salmon still challenges boatmen with powerful currents, waterfalls, and boiling rapids. Below, the sparkling Middle Fork of the Salmon meanders through a narrow valley. The Middle Fork flows 104 miles to its mouth at the Salmon, alternating quiet stretches with rushing water. Tucked deep into the northwestern U. S., it retains its primitive character under protection of the Wild and Scenic Rivers Act of 1968; the Salmon, one of the longest undammed rivers in the country, may one day receive similar protection.

WILLIAM R. GRAY, NATIONAL GEOGRAPHIC STAFF; ED COOPER PHOTO (OPPOSITE)

Forested slope of Pine Creek Gorge nudges
its namesake stream in north central Pennsylvania.
ConRail tracks follow the curve through stands
of pine, hemlock, maple, and birch.

The East:

Pine Creek Gorge

Photographed by William S. Weems

Canoeists struggle in vain to stay afloat in a Pine Creek rapid. Nearby, in a quiet stretch, a fisherman angles for trout and bass. At lower right, a hiker nears the summit of a hill on the Black Forest Trail, not far from the creek. Picnickers, campers, and hunters also seek recreation and relaxation in Pine Creek Gorge, a 50-mile-long rift in the Appalachian Plateau. In 1968 Congress included Pine Creek in a group of streams deserving consideration as Wild and Scenic Rivers, and the National Park Service named the gorge a Registered National Natural Landmark. Nearly 10,000 boaters floated Pine Creek in 1978. This couple—though not wearing life jackets—made it safely to land, losing little but their dignity.

SALLIE M. GREENWOOD

41

Nameless tributary tumbles down a mossy ravine
toward Pine Creek. Fungi cling to a fallen log
that bridges the rivulet. Along its banks, and throughout
Pine Creek Gorge, wild plants and animals thrive.
A tiger swallowtail butterfly alights on a lilac
(below, left), and a hay-scented fern spirals upward
toward the sunlight. Oldest of Pine Creek's flora,
horsetail, or *Equisetum* (above), grew when dinosaurs
roamed the earth, and has remained virtually unchanged
for 300 million years. Here author Ralph Gray
(at left) and Peter Stifel, a professor of geology at the
University of Maryland, examine one of the wispy plants.
Settlers called them "scouring rushes" for their
abrasive stalks, and used them to clean pots and pans.

By
Ralph Gray

Cañon. Spaniards brought the word and the concept to the American Southwest. English-speaking Americans added the "y" and carried the term all over the country, even to the East. Spectacular breaks in the land that had existed for generations in the public imagination as gorges became canyons.

Such a one is Pine Creek Gorge, sometimes called the "Grand Canyon of Pennsylvania." For 50 miles it cuts a cleft as much as 1,000 feet deep through the Appalachian Plateau of north central Pennsylvania. Some 350,000 people a year come to this unspoiled corner of "Penn's Woods" as active recreationists—canoeists, rafters, fishermen, hunters, hikers, lovers of flora and fauna.

Calling a gorge a canyon does not change its looks. Eastern canyons remain wooded and softly molded. They offer moments of leafy intimacy not provided by square miles of bare, glaring rock. But don't be lulled to sleep, either. Eastern canyons boast hanging precipices, soaring rock outcrops, big timber, booming waterfalls, and roaring rivers still cutting deep into ancient rock.

And everywhere, except in the southern states, you can hear echoes of the gnashing and grinding of Ice Age glaciers. The ice sheets created, shaped, redirected, or actually reversed drainage patterns of the East, giving its canyons a character found nowhere else in the United States.

The long-gone glacier was an ever-present phantom companion during a recent holiday weekend canoe trip through Pine Creek Gorge. Where we put in at Watrous, the river-size creek coursed east through a fairly open valley. Soon the great wall of Mount Tom rose before us, seemingly blocking the waterway. Where would the river go?

At the last moment, as our canoes hurtled toward the barrier, we saw the river's escape route. To the northeast a broad valley opened up. But no, Pine Creek turned abruptly south, digging and plowing its way through a massive plateau.

"It's the work of our old friend, the glacier," Peter Stifel shouted to me from his canoe. "At one time Pine Creek did flow to the northeast and into the Tioga River. But the glacier blocked that route and forced it to find a new exit—south to the Susquehanna River."

Later Peter, a member of the geology faculty at the University of Maryland, came up to me on the shore. Between thumb and forefinger he held a feathery green plant that most people would dismiss as an undistinguished weed. "This plant witnessed the glacier's work," said Peter. "It's called *Equisetum*, or horsetail, and it's ancient. Its early relatives, which were almost identical to it, were preserved as fossils in the canyon's sandstone rocks 300 million years ago."

This was my first canoe cruise with Peter, but many of the 17 men with me had enjoyed almost ritualistic spring rendezvous on rivers of the middle Appalachians each Memorial Day weekend and at various other times for 30 years or more. We were a floating group in more ways than one, with a core of old friends and relatives, plus new people asking in from time to time, and others dropping out in favor of such questionable alternatives as sailing, golfing, or staying home with their families.

Pine Creek was a favorite trip, which we had made twice before. We were delighted in 1968 when Congress passed the National Wild and Scenic

Blanket of green wraps Pine Creek Gorge, nearly denuded by loggers in the late 1800's. Lumber towns grew and died here, and narrow-gauge railroads, their tracks now pulled up, left a network of grassy trails.

Rivers Act. It included Pine Creek in the category for further study. Subsequent statements by the Heritage Conservation and Recreation Service have recommended the creek be named a National Scenic River because it possesses the outstanding scenic and recreational qualities and the abundant fish and wildlife that the law requires. Also, the level and quality of its water ensure boaters what's called a "meaningful recreational experience." Pennsylvania is also considering making Pine Creek a part of the state's Scenic Rivers System.

On this trip we would see for ourselves. We had the support of Ed McCarthy, an outfitter known far and wide as "king of the canyon," and a supporter of its protection. Ed turned out to be one of the area's natural wonders himself. Well into his seventies, he is hard as rock and leads many trips by canoe or rubber raft through the gorge during the boating season.

It was almost as if Ed actually operated Pine Creek on a timetable. When we put in at Watrous, he looked at his watch and announced, "We'll camp at the mouth of Darling Run, arriving at 4:35 p.m." Sure enough, at 4:30, after making the right turn at the foot of Mount Tom, we saw the Darling Run campsite come into view.

Next morning, after a perfect night under the stars, and with one of schoolteacher Merle Dusek's monumental breakfasts under our belts, we pushed off into the depths of the canyon.

"Soon we'll come to Owassee Rapids," Ed boomed in the general direction of our nine canoes. "It's the most dangerous spot on the river. The water piles up on the right and tumbles down through a series of standing waves. If you're brave enough, or good enough, or foolish enough, you can follow the current to the right. But my advice is: chicken left!"

We chickened left, every canoe coursing the bounding waters without a scrape. But Pine Creek rose up to smite a craft following our party, rudely dumping its occupants—including the family dog—into the water.

Now we were in the deepest section of Pine Creek Gorge, looking up hundreds of feet at tiny people standing behind the guardrails of overlooks in Colton Point and Leonard Harrison state parks. Ed continued his travelogue as we moved along on the swift-flowing and often noisy water. "In 11 minutes we'll see Tumbling Run, a stream that splashes down a glen cut out of the right canyon wall." Arriving on schedule, we admired the falling waters at a moment when they were fractured into crystals by a backlighting sun.

The primeval scene filled me anew with an old Pine Creek feeling. It was a feeling of the empty woods, the wilderness, the frontier still existing all around me in a time warp that made Colonial times only a little farther off than World War II. The Indians had just left, and deer and bear were still watching from their hiding places.

And Ed McCarthy was a modern-day Nessmuk, lover of the wild and extoller of its virtues. Nessmuk was the pen name, borrowed from a young Indian friend, used by George Washington Sears in signing his *Forest Runes* and his book on woodlore. In 1884 Nessmuk published his classic *Woodcraft*, the first book on outdoor life to appear in the United States. Though often associated with the Adirondacks, Sears spent most of his adult life in the Wellsboro area of Pennsylvania, where much of his material grew out of countless rambles in Pine Creek Gorge.

A literate Natty Bumppo, Sears was a forerunner among writers who awakened America to an appreciation of its wilderness heritage. Many of his early poems were written on birchbark.

We stopped for lunch at Tiadaghton, Iroquois for "river of pines." It

Pine Creek Gorge, the "Grand Canyon of Pennsylvania," spreads wooded fingers into the Appalachian Mountains. Most of the gorge lies within state parks and forests.

is the only place in the gorge where a road winds down from the rim to the water's edge. It was a Sunday, and the park was swamped with picnickers and their aluminum campers. Ed took in the sandy beach, the green-grass amphitheater backed by large hardwoods, and the canyon walls, and sadly said, "You can see what the Indians enjoyed, what we have now, and what could be again."

One of the friends I had invited to join our group was Arthur P. Miller, Jr., of the Philadelphia office of the National Park Service. "Could the Park Service help protect Pine Creek Gorge?" I asked Art.

"We have no plans to give Pine Creek any kind of National Park designation—as we do with New River in West Virginia," Art said, "but we are ready to support the state in its efforts to preserve the gorge and protect it from misuse. Pine Creek Gorge is also established as a Registered National Natural Landmark. That means that the Pennsylvania Department of Environmental Resources is empowered to preserve the area as it is."

Soon the hamlet of Blackwell came into view on the left, with a highway angling down from the outside world and following the creek for the rest of its course. Again, mobs of people with rafts and canoes and leftover picnic materials crowded the shore where Ed had said we would camp. However, they were day-people, not campers, and soon headed back to civilization, leaving the area to us.

Here we said goodbye to Ed, who had to shepherd some clients by car back to his Antlers Inn headquarters. Rather, Ed said goodbye to us, for we had learned early that talking with Ed was more like listening to a Shakespearean monologuist. Stepping up to an invisible podium, he addressed us for the last time, answering his own questions.

"You say you might need more firewood? If so, send a brace of your young stalwarts to yonder tree and chop away. Why that particular tree? Because, like Methuselah and Abraham Lincoln, it's dead."

This tree had died a natural death, but most of the original forest cover of the gorge and surrounding areas died by the logger's hand. In the late 1800's this upper Susquehanna region became the logging center of the nation. The blanket of virgin pine that covered the gorge gradually disappeared, to be floated down Pine Creek each spring in rafts of logs.

Little remains to be seen of the days when big timber was a way of life hereabouts. Nature in its forgiving way has reforested the cutover lands. But once, on the West Branch of the Susquehanna, we did find a memento of Pennsylvania's logging heyday. A steel ring looped through the eye of a spike was embedded in a deeply anchored boulder on the left bank. We knew it to be the fastening for a cable boom that once stretched across the river. Such booms held acres of logs waiting to be formed into rafts for the float to Harrisburg.

Paddling for two more days down the still-swift lower Pine Creek, we continued another tradition—keeping alive the silent, undeclared war between wading fishermen and canoeists. I have never seen this war erupt into anything more overt than sudden glares, quickly masked by the realization that the other guy has as much right to fish in the best hole immediately below a riffle as you have to canoe through the only spot possible without scraping. But it does add to one's canoeing skills, amid tumbling white water and grasping boulders, to be on the lookout for a thin nylon line with a submerged fishhook.

In terms of modern-day recreation, fishermen no doubt discovered Pine Creek Gorge before canoeists did. Theodore Roosevelt, with his quenchless thirst for primitive beauty, is known to have come here to fish and enjoy the solitude.

"Was that really before you and Dad started canoeing?" asked my irreverent nephew Chet, referring to me and my older brother Harold.

It is true that Harold and I, the senior members of this group, have been canoeing so long that our hair, especially mine, has turned almost the color of the water we love the best. That night in camp, Chet's brother Tom said, "I do sometimes wonder how long you and Dad will keep going on these canoe cruises, and what it is that brings you back year after year."

Teddy Roosevelt again came to mind. In his mid-fifties, he had gone on an ill-fated expedition up the "River of Doubt" in Brazil. He contracted a tropical disease from which he never fully recovered. When asked why he had made such a risky trip at his age, he said, "I had just one more chance to be a boy, and I took it!"

Harold and I looked into the glowing campfire. We felt the eternal bond of the wilderness, and heard the stream swiftly slipping by in the dark. Each of these trips, we knew, might be our last chance to be boys.

Niagara Gorge

It's hard to go to Niagara and turn your back on the falls. But when your objective is Niagara Gorge, Niagara Falls becomes merely the upper anchor of your study. To maintain the proper orientation I kept in mind Mark Twain's comments on the power of imagination—how one's preconception of a wonder like Niagara Falls builds a spectacle in the mind to which the real falls is inevitably "a poor dribbling thing."

And so it was that Jack Krajewski and I stood on Prospect Point, at the very brink of American Falls, and looked north down the length of a deep, rock-walled gorge that is truly remarkable in its own right.

More than 200 feet deep, Niagara Gorge runs for seven miles between the falls and the Niagara Escarpment, marking the boundary between the United States and Canada. It is almost totally the creation of Ice Age glaciation. "Let's go down to the lower end of the gorge, and you'll see how the canyon and the falls came to be," said Jack, Senior Scientist in geology for the Schoellkopf Geological Museum, part of the Niagara Frontier State Parks Commission.

"This is the face of the Niagara Escarpment," said Jack a few minutes later as we stood on a woodsy ridge that marched across the New York-Ontario countryside, broken only by a gap where the river flowed through. "This is where Niagara Falls was born 12,000 years ago." He explained that as the last ice sheet retreated, some of its meltwater formed the ancestral Great Lakes. One of these lapped against the Niagara Escarpment and began pouring over it at a number of different outlets. Eventually, the largest and westernmost of these spillways became the Niagara River, punctuated by Niagara Falls where it tumbled over the escarpment. The drainage from the upper Great Lakes rushed through this one chute with such force and volume that the falls ate their way backward to their present location at the rate of three or four feet a year.

So much for the overview; now to get down into the gorge. From the rim on the New York side, two trails drop down to the bottom. Just above water level, an abandoned trolley line forms a walkway following the river for miles. Jack and I rock-hopped to the very edge of Whirlpool Rapids, a boiling maelstrom in the Niagara River. The fascinating world of nature in the canyon walls above seemed far removed from both the crashing waters and the industrial city on the rim. A surprising variety of trees and plants cloak the walls, haven for birds and ground creatures.

My final descent into the gorge was by elevator from Prospect Point. I put on waterproof gear and stepped aboard *Maid of the Mist* for the exciting voyage to the extreme upper end of the gorge. We ventured so close to the foot of the falls that all consciousness was reduced to a roaring whiteout of mist and spray. Looking directly into the face of Mark Twain's "poor dribbling thing," I had to admit that my imagination had never construed anything approaching this.

New River Gorge

Probably the most concentrated dose of white water in the United States, New River Gorge in West Virginia counts 21 major rapids in one 15-mile stretch. But the New River does not rest its reputation on white water alone. Geologists respect it for quite a different reason. Despite its name, it is the oldest river in North America.

New River is the headwater portion of the ancient Teays River, which a hundred million years ago began tumbling out of the ancestral Appalachians, once mightier than the Rockies. It had its source in what is now the northwestern portion of North Carolina. It flowed north (another oddity) through Virginia and into West Virginia, then angled through Ohio, Indiana, and Illinois, where the Mississippi joined it. The last glacier obliterated the western half of this thousand-mile-long stream, throwing a dam of ice hundreds of feet high across its course in south central Ohio. From this point to its source, the Teays River—now called the New—was untouched by the glaciers, and maintained its ancient course.

And what a course it is! The final 66-mile section, from Hinton to the mouth, cuts valleys and canyons a thousand feet deep into the Appalachian Plateau. The farther north it goes, the deeper it gets. At Thurmond the waters narrow and enter the wildest, most remote part of New River Gorge. From here to Fayette Station only one person lives along the river, where once a string of coal towns flourished. No road follows the waters. The only accesses are the river itself and the Chesapeake and Ohio Railroad.

My daughter Donna, a West Virginia schoolteacher, joined me in sampling the natural and man-made charm of the area. We had driven through nondescript hills to arrive at Thurmond on a one-lane wooden afterthought of a bridge tacked onto a railroad span. The road ended, and we walked along "Main Street," a narrow sidewalk between the C&O tracks and a ghostly line of abandoned brick and stone buildings. One was once a bank, but now it offers shelter to wayfarers as the Bankers Club. Erskine Pugh, the proprietor and restorer of the building, was our constantly visible, gracious host, and also mayor of Thurmond, population 87.

During the coal boom that hit New River Gorge in the early 1900's, more than 500 people lived in Thurmond. It was the throbbing hotel, entertainment, and banking center for the area. Newly rich coal mine owners, businessmen, gamblers, and adventurers of both sexes made Thurmond the "Dodge City of the East." Legend has it that a poker game in one of the hotels lasted 14 years.

Erskine Pugh awakened us at dawn, saw that we had a good breakfast, and pointed us toward Jon Dragan, the energetic entrepreneur who heads up Wildwater Expeditions Unlimited. Jon put us on one of his rafts with eight other passengers and gave us one of the best river trips ever.

As we plunged deeper into the canyon, the rapids became larger, more frequent, and wilder. Jon and his young boatmen, Dave Gill and Peter Grubb, guided us expertly through them all, including some neck snappers that are rated class five on a scale of six. (Six is well-nigh unrunnable.) Near the end, already sopping wet from flying spray, we shot through three monstrous rapids with a combined drop of 32 feet in less than a quarter of a mile—a run that reminded me of Crystal Rapids, in Grand Canyon.

Landing, Donna summed it all up. "It's amazing how few people, even in West Virginia, really know about the New. Maybe, being the oldest river in North America, it's a river whose time has finally come."

The glaciers had nothing to do with the formation of Linville Gorge. This 2,000-foot-deep chasm on the east slope of the Blue Ridge in western North Carolina was quite simply carved out by the Linville River in its pell-mell rush to the Piedmont. Deepest of all eastern gorges, Linville is rimmed on the east by Jonas Ridge and on the west by Linville Mountain. Both enclosing arms reach elevations of 4,000 feet. In 1951, the 7,600 acres of the gorge were set aside as a "wild area," the first such preservation east of the Mississippi. As at Niagara, Linville Gorge begins below a waterfall—Linville Falls, a deeply clefted drop of 102 feet.

Though it was mid-July when I visited this scenic wilderness, *Rhododendron catawbiense* was still in flower, but past its prime. With Ranger Les Burril I walked down Pine Gap Trail one morning through a botanist's paradise of laurel (as local mountaineers call rhododendron), true mountain laurel, alder, and huge virgin hemlock.

At Wisemans View we met Les's colleague, C. W. Smith, a Forest Service technician, who was supervising a crew putting in a new asphalt walk from a parking area to the brink. "I sometimes come up here in the evening," said C. W., "and watch the Brown Mountain Lights." I studied his eyes. He wasn't kidding; he *had* seen "the lights," an eerie, unexplained luminescence that seems to play in the sky above Brown Mountain. Unexplained is hardly the word; a score of explanations have been offered, ranging all the way from pure superstition to auto headlights refracted upward. No single theory satisfies everyone.

We came back to steadier ground when C. W. started pointing out landmarks as solid as rocks. Several of them, in fact, were rocks—massive outcroppings on the east rim of the gorge. From these granite skyhooks hangs a mottled green drapery of virgin forests and woods. "In all this huge gorge," C. W. said, "only two little coves on the east side were ever logged. All the rest is the way God made it." Even the 28 miles of trails are only pin scratches in the fabric. From ten feet away, they are invisible.

I followed a trail on the east rim to the summit of Tablerock, a great flat-topped granite monolith, and it seemed that all the world was at my feet, or within reach of my eyes. East lay Lake James at the foot of the gorge. Beyond it, however, the cities, farms, and the factories of the heavily populated Piedmont dwindled to a hazy horizon.

But west, in the mountain world, the land was bright. A front had gone through the previous night, washing away the shroud of civilization. The crystal clear air sharply outlined ridge after ridge to infinity. From where I stood I could see other peaks that I had climbed in other days, and they spoke to me across the distances—Mitchell, the highest east of the Mississippi; Roan, on the Tennessee border; Grandfather, far to the north. I wondered: Why did this glorious scene seem so unusual? Wasn't this the way the world was supposed to be?

Sundering a wall of granite, Linville Falls spawns mist that waters tenacious rhododendron and virgin hemlock. The river has carved Linville Gorge, 12 miles long and 2,000 feet deep, in western North Carolina. Walls too steep for logging have preserved the gorge's wilderness.

JOHN EARL

Niagara Gorge

"This wonderful Downfal . . ." wrote a Belgian-born missionary of Niagara
Falls in 1678. "The Waters . . . do foam and boyl." They also gradually wear away their
sedimentary bedrock, the Niagara Escarpment, as they tumble over it. In the last 12,000
years, the falls have retreated seven miles, leaving a gorge 200 feet deep
downstream. They continue to move three or four feet a year, and within 50,000 years
will probably disappear. At left, nighttime floodlights tint American Falls.

New River Gorge

Stairstepping downward, the New River works its way northward from the eastern
slopes of the Appalachian Mountains, one of the few rivers to do so. Its gorge, in the wooded
hills of West Virginia, extends 66 miles and averages 1,000 feet deep. In 1978 Congress
named the New a National River, freezing development and prohibiting change. Along its
banks, hikers may stroll beside carpets of trillium (left), gaze into constellations
of bluets, or surprise a spotted fawn. Downstream from here the gorge deepens and the
river narrows, offering some of the East's most challenging white-water rafting.

The Southwest:

"Grandest of God's terrestrial cities," naturalist John Muir called the Grand Canyon.

Grand Canyon

Glowing afternoon light whitens peaks along the North Rim.

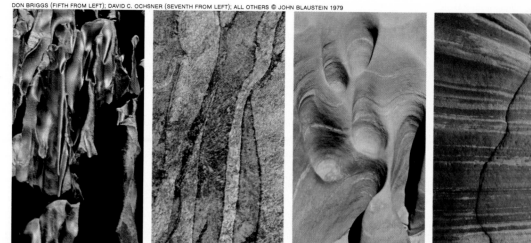

SCHIST   SCHIST-GRANITE   DOX SANDSTONE   TAPEATS SANDSTONE

Rocky pages in time, walls of the Grand Canyon
tell a story nearly two billion years long. A ridge
of Wotans Throne (right) discloses a sequence of
building and eroding, of heating and cooling, of
rivers and seas, and of desert winds. The most
ancient canyon rock, Vishnu Schist (above) appears
smooth and massive. Other formations show the
variety of nature's artistry (top, left to right):
two schists, one polished to a silky sheen, the other
seamed with granite. Dox Sandstone acquired its
sinuous undulations in a tidal-zone delta a billion
years ago. Formed near a seashore, the rust-red
Tapeats came 500 million years later, and movement
in the earth formed the sharp angles of the Supai.
Lava flows created the basalt pillars. Slate-
textured Coconino Sandstone predates the basalt,
and contains fossil tracks of desert creatures,
such as scorpions. The coarse sand this one
scurries across provides raw material for new
episodes in an endless geological saga.

SUPAI   BASALT   COCONINO SANDSTONE   COARSE SAND

"The waves . . . completely submerged me and almost took my breath away," wrote an early boatman on the Colorado River. Today, adventurers like the one at upper left share the same experience. Martin Litton (left) guides groups through the Grand Canyon in wooden dories, running rapids in what some call the roughest navigable river in the world. The beauty and isolation, as well as the thrills, bring thousands into the canyon to experience its many faces—from the dark force of a thunderstorm between steep chasm walls (above) to the quiet fragility of rain-drenched phlox brightening a rocky ledge (right).

61

Showcase of flora and fauna, the Grand Canyon exhibits
a complex mosaic of life forms. Deep in the canyon,
the slow-growing agave, or century plant (left), may take from
12 to 20 years to flower, and will die soon after.
An ascent from the canyon floor to the North Rim, like
a journey from Mexico to Canada in one vertical mile,
moves from scorched desert to damp forests.
Moisture-frugal prickly pear and barrel cactus (below)
give way to fir and spruce at the rim. A mule deer fawn
(right), motionless and alert, pauses from browsing.

Preceding pages: Piñon trees weighted with new-fallen snow frame the South Rim's Yavapai Point. The oldest evidence of man in the Grand Canyon dates from 4,000 years ago. In those days, only Indian hunters ventured to the "big water deep down in earth." The first white man to see the canyon, a Spanish captain named García López de Cárdenas, reached it in 1540. It remained largely unexplored until 1869, when John Wesley Powell, with the avowed aim of adding a "mite to the great sum of human knowledge," set forth "down the Great Unknown." The intrepid one-armed Civil War veteran led two grueling expeditions through the canyon, usually perched on a chair lashed to the deck of his boat.

Paradise found, Elves Chasm (opposite) beckons with clear water and verdant growth only a few feet above the Colorado. Level upon level, the chasm and its splashing cascades lure climbers. Downriver, near Havasu Creek, ferns and mosses watered by a seeping spring weave a tapestry high on a canyon wall. Many such miniature worlds lie hidden in side canyons along the river. These small havens renew perspective for eyes that have lost all sense of scale in the vastness of the Grand Canyon. A columbine (right), peeking from a dark crevice near the river, takes on the color of fire in the sun.

"Mudslingers" in flooded Kanab Creek pause after smearing themselves with brown silt—a frolic that has become a tradition with boaters in the Grand Canyon. Once dismissed by a 19th-century explorer, Army Lt. Joseph Ives, as a "profitless locality" to remain "forever unvisited and undisturbed," the Grand Canyon now draws three million visitors annually. Fragile trails and scarce campsites have become overburdened. Thousands of river runners vie for reservations on boat trips that as recently as 1953 only 200 had made. With the crowds come litter, overuse, and an end to the tranquillity they seek. Conservationists strive, as one puts it, to "leave nothing behind but paint scrapes on the rocks from our boats."

Gentle-eyed burros, introduced to the canyon by prospectors, thrive and multiply—and provoke an ecological controversy. Some experts want them relocated because they tear up slow-growing desert vegetation, compete for food with native animals, and damage archaeological sites. At top, phalaropes in winter plumage wing close to the river in a rare canyon appearance during migration.

By
Thomas
O'Neill

The time is late 20th century, but man, it seems, has yet to walk upon the earth. As light spills from the sky at daybreak, colossal forms emerge from the dark, islands of rock afloat on morning's shadows. Ribbons of color—gold, crimson, purple—paint the coarse walls. Day advances, and shapes multiply in variety; a stretch of green appears below, a river cutting through rock. The air is immaculate, the stillness profound. Rock. Water. Light. All is elemental, all is untouched.

The sun climbs higher, and the world changes. Unexpectedly, a family of hikers puffs by; boats string out on the distant Colorado River; a tour bus wheezes to a stop; a sight-seeing plane drones overhead. The world has been peopled. Yet for a prodigious moment, usually at dawn on the rim, the endangered fantasy of a world without man lives here in the Grand Canyon, perhaps the most spectacular and overwhelming place on the earth's surface.

It is a chasm 277 miles long, as much as 18 miles wide, and a mile deep, all sculpted by erosion, located in northwestern Arizona. People do come. Three million visitors crowd the rims each year, making Grand Canyon National Park, established in 1919, one of the most heavily visited parks in the country. Like a great epic novel, the canyon burgeons with tales—subplots of geology and wildlife, river running and conservation, Indians and explorers. Passage through its narrative is one of the greatest journeys we can take.

A river provides the best vantage for understanding the canyons of the Southwest, I decided. So on a baking July day I pushed off from Lees Ferry, Mile 0 on the river, at the northeastern corner of the park, and began a 16-day journey down the Colorado. My companions were young biologists from the Museum of Northern Arizona in Flagstaff. "We're going to be surveying the native fishes in the river," explained George Ruffner, the trip leader, as he put his back into rowing a large raft.

It is an important study, one that confronts the question of how the modern world's intrusion on Grand Canyon—in this case, Glen Canyon Dam—has affected life between the rims. One conclusion is already apparent: Several species of native fish are disappearing.

Before 1963, when the dam was completed above Lees Ferry, water temperatures fluctuated between freezing in winter and 80° F., in summer. Today the water, released from the bottom of dam-impounded Lake Powell, is always below 50° F., making the upper river too chilly for humans to swim in, and apparently too cold for such indigenous fish species as Colorado River squawfish, humpback chub, and razorback suckers to spawn in. Introduced game fish, such as rainbow, brook, and brown trout, aggressive predators, also threaten native species.

We stopped at the major warm-water tributaries, where the native fish tend to congregate, and took samples both by setting nets and, new to me, shocking the fish. A generator sent a small charge through the water, and within a minute stunned fish, mostly flannelmouthed suckers, floated to the surface, lying like lily pads waiting to be netted. We measured, weighed, sexed, and tagged the unharmed fish before throwing them back into the river.

Afternoon light casts a dusty purple haze over Chuar Butte on the North Rim. The Grand Canyon, called by Theodore Roosevelt "one of the great sights which every American . . . should see," continues to stagger visitors with its monumental proportions. Adventurers explore its countless cliffs and side canyons, and seek out still-untouched wilderness recesses.

Our longest stop was at the Little Colorado River, major habitat of the humpback chub, a fish that once swam freely in the Colorado but has been on the endangered species list since 1967. "The Little Colorado is too warm for the predatory rainbows," said biologist Chuck Minckley, "and the Park Service has banned camping and fishing within half a mile of the mouth." Soon after we tied up, chub began gathering in the shade of the rafts. We set the nets and spent the afternoon gingerly removing the fish, tagging 27 altogether. It is a queer-looking fish, this silvery member of the minnow family. *Gila cypha* sometimes grows longer than 17 inches, and behind its head is a small knob of muscle tissue, like a topknot. "It's like holding a bald eagle," mused Wendell Minckley, Chuck's brother, as he lifted one of the fish. "They're on the same list."

I left the nets, fish buckets, and biologists at the Little Colorado and switched from a raft to a party of wooden dories. Able to hold four or more passengers and an oarsman, the dories somewhat resemble the Grand Banks fishing boats, with their upturned prows and high, flaring sides. In the water they are stable and maneuverable—the better to avoid wood-splintering rocks—and no craft rides the rapids with more grace.

The Colorado is probably the greatest white-water river in the world. Some 160 rapids churn the Grand Canyon section. On my first two days in the dory we hit the wildest stretch. First came Hance Rapids. Water boiled around the sides of the dories, the noise drowning thought, but we caught a good path and rollicked with the waves, blunting breakers with the soaring bow. At Sockdolager we raced between two encroaching, boat-eating waves. Luck fled at Horn. A large wave hung over the boat like an imposed sentence before crashing over the gunwales, tearing an oar away, but we wobbled through, marinated in the cold Colorado water. Hermit, with its deep troughs, provided a swooping, soaring ride.

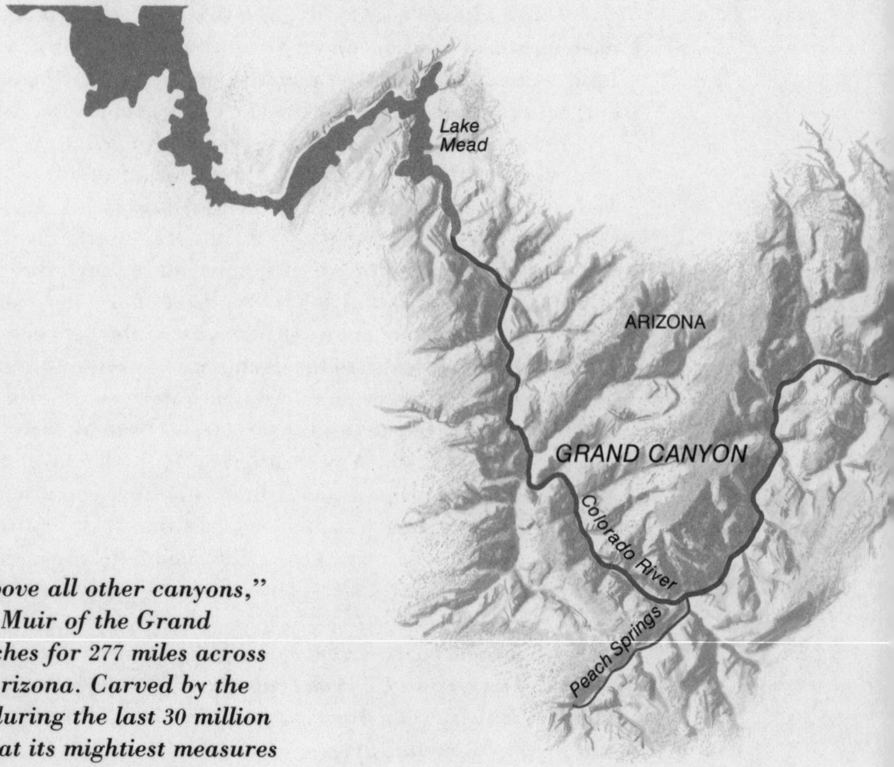

Lake
Mead

ARIZONA

GRAND CANYON

Colorado River

Peach Springs

"Supreme . . . above all other canyons,"
wrote John Muir of the Grand
Canyon. It stretches for 277 miles across
northwestern Arizona. Carved by the
Colorado River during the last 30 million
years, the chasm at its mightiest measures
18 miles across and a mile deep.

Few people run the Colorado without giving thought, and perhaps benediction, to John Wesley Powell. The Grand Canyon had been little more than sketched on maps in 1869 when the one-armed Civil War veteran and eight men in small wooden boats successfully challenged the Colorado River in a canyon he called the "Great Unknown." His account chronicles hunger, desertion, and a constant war with the "mad waters." On August 28, three men left the party in a futile attempt to climb to the rim; on August 29, after 26 days, Powell and the other half-starved men rowed out. Powell later wrote: "The first hours of convalescent freedom seem rich recompense for all pain and gloom and terror."

In addition to his stirring tales of adventure, Powell brought back the first informed description of the Grand Canyon's geology. Unlike Powell, who was preoccupied with saving his skin, the modern river runner has plenty of time to read the rocks. "Grand Canyon is primarily a geologic park," contends William Breed, head of the geology department at the Museum of Northern Arizona. "There is not a place on earth where geology is revealed on such a fantastic scale and in so much detail."

During my trip in the dory I found myself on an immense journey in time, humbly passing through more than a billion years of the earth's history. Near Lees Ferry the cutting river reveals the yellow-gray Kaibab Limestone, deposited as sediment 250 million years ago, before mammals or flowering plants existed. Fossils of coral and early sharks are encased in the rocks. Below it lies another limestone, named Toroweap, and then Coconino Sandstone, the solidified remains of sand dunes laid down in a Sahara-like desert. The red rocks of Hermit Shale and the Supai Group follow—indicators of a savanna-like environment—and at Mile 22.6 the imposing Redwall Limestone appears, the

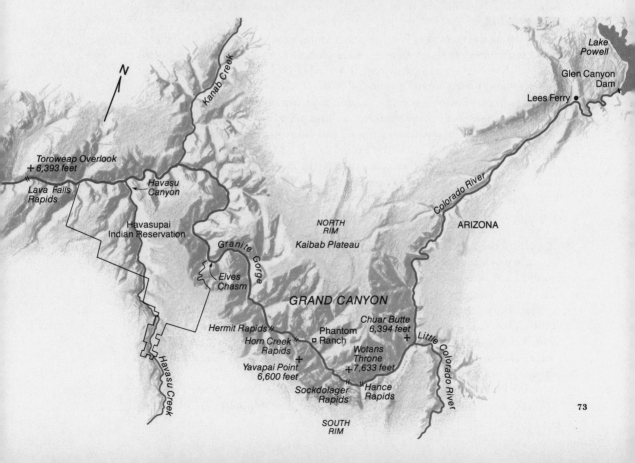

dominant cliff-forming rock in the canyon. Numerous marine fossils in this limestone give proof of a deep, quiet sea 330 to 340 million years ago.

The river unveils another eight layers of rock as we travel deeper into the earth. At Mile 78 we enter a dark and narrow chasm, known as Granite Gorge, the appropriately portentous setting for the oldest rocks in the canyon—the Vishnu Schists, metamorphic rocks 1.7 billion years old. The sheer, coal-black walls, formed before complex species of life existed, are the roots of an ancient mountain range. The canyon is a mile deep at this point, and the Kaibab Limestone, once at my elbow, now forms the canyon rim high overhead.

The chief architect of canyons is water, usually a tumbling, surging stream. Sometime between five and thirty million years ago the Colorado River, in a manner still disputed by geologists, began to force its way across the Kaibab Plateau in the southwestern U. S. With its grinding load of gravel and sand, it has carved a monumental trough. The cutting persists today, at the rate of a few inches every thousand years.

Meanwhile, torrential seasonal floods from the rim, plus rockslides and weathering, have excavated a vast system of side canyons. An awesome bare-rock topography of butte and pinnacle, amphitheater and temple, has formed between the rims. "We are fortunate to be on earth at the time when the area has been uplifted and eroded into a canyon," concluded Bill Breed. "Remember, for most of its history this region was at or near sea level—not a very spectacular place."

Boat passengers usually glimpse their terrestrial counterparts, hikers, in the vicinity of Phantom Ranch, at Mile 88. Nearby, the park's three maintained trails from the rim to the river come down to the Colorado. In summer as many as 1,500 people a day troop down the twisting paths. A smaller number ride mules, those imperturbable beasts that seem unconsciously to delight in swinging wide-eyed passengers along the very edges of cliffs.

Just as the river trip is a journey through time, the hike from rim to river is a journey through climate. To walk 6,000 feet down from the North Rim to the river is the equivalent, in terms of climate, of traveling 3,000 miles from central Canada to central Mexico. A forest of ponderosa pine, Douglas fir, and aspen gives way to twisted piñon and juniper pines, and finally to a desert of agave, blackbrush, and prickly pear cactus.

The first white men to see the Grand Canyon were Spanish soldiers searching for cities of gold in 1540. They found instead an impossible barrier. Three centuries later, in 1858, with the region still untried and unknown, Lt. Joseph C. Ives entered the "Big cañon" with a small exploratory expedition. In his report he described the canyon as a "profitless locality," and darkly added, "It seems intended by nature that the Colorado river . . . shall be forever unvisited and undisturbed."

Lieutenant Ives was no prophet. But how could he have foreseen that in little more than a hundred years a craving for "profitless" wilderness would send swarms of people boating and rafting through the Grand Canyon? In 1972 alone, 16,432 people ran the Colorado, more in one year than in the entire century following Powell's 1869 trip. Alarmed by this catapulting increase, the Park Service froze the number of river passengers and inaugurated a five-year study of the river environment. The findings were sobering: Serious, perhaps irreparable, damage was being inflicted along the Colorado corridor.

In the days when the river ran free, seasonal floods swept detritus from the beaches. Glen Canyon Dam, however, has subdued the Colorado, drastically reducing the floods. Consequently, human fecal waste, amounting to 20 tons a year, and charcoal left from wood fires were piling up on beaches, creating health and aesthetic hazards. Foot trails to the popular side canyons were destroying vegetation and advancing erosion.

The Park Service's response came in a 1977 draft management plan which required that all fecal waste be containerized and hauled out of the canyon. No wood fires are allowed during the summer; only single trails to the side canyons are permitted; and a river monitoring program has become permanent. Rangers say that after just one year they could see an improvement in the condition of the beaches.

The plan also addressed that intangible something called a "wilderness experience." To park administrators it implies an intimate and edifying river passage free of congestion and noise, something the Colorado during its summer glut of fast, noisy motor trips does not always offer. To honor that definition and encourage trips that are longer and more exploratory, the plan calls for lengthening the summer season, instituting a winter season, and phasing out all motors from the river. "We're establishing game rules, just as in tennis or baseball," says David Ochsner, Chief of Resource Management. "Only so many people can occupy the field, and certain rules, such as taking out trash, have to be obeyed."

The Grand Canyon's only longtime residents live up a side canyon called Havasu. They are the Havasupai Indians, perhaps the smallest and most isolated tribe in the country. Their ancestors settled in the canyon around 1300. I entered the reservation by way of the rim, 67 miles northeast of Peach Springs, riding a horse into a narrow, deepening, red-walled canyon until I came upon a lush pocket of vegetation where the village stands. It has been home, garden, and prison to the Havasupai since 1882, when the federal government assigned the tribe this niche of 518 acres.

Brightly painted prefabricated houses looked out of place as they anchored tiny irrigated fields of alfalfa and hay. On the dirt streets young Indians gallop their horses, while elders linger in the shade of the post office and grocery store. All seem oblivious to the steady trickle of backpackers who have hiked in for a look at the waterfalls of Havasu Creek.

The population of the reservation stands at 476, up from 300 in fewer than ten years.

"When you're living in a remote place, there's not much to do but increase population," said Wayne Sinyella, the 30-year-old tribal chairman. January 3 is a holiday on the reservation, he told me. "On this day in 1975 the park-expansion bill granted us 185,000 acres of land on the rim, 84,000 of it from the park." Handed this elbow room, the Indians plan to use it primarily for grazing livestock. Environmentalists objected to the agreement, fearing the tribe would sully the land with development.

"No," said Sinyella. "We don't want to see the place as a big town. We want to live up here peacefully with a very few cabins . . . just like in the old times."

The Indians have never used the river much, except for irrigation and swimming. Only the foolhardy newcomers, we tourists, feel compelled to run Lava Falls, the world's fastest navigable rapids. Wally Rist, head boatman, dutifully tried scare tactics. "Have you ever watched clothes in a dryer? That's what it will be like." From a mile away we heard the roar, a deep thundering, the sound of dynamos. Half the group took Wally's cue and hiked around the falls; the rest of us went into the maw. My heart added its own noise as our dory, the lead one, slipped off the glassy tongue into the rabid water. I knew what route we wanted to take, but all stratagem seemed lost in the great gnashing of waves. The water spun us like driftwood, slapping us over and over, but after 30 intense seconds our punch-drunk boat emerged right side up. The rest of the fleet followed.

Days later I stood on Toroweap Overlook, where only a minute fraction of the park's visitors venture, and gazed 3,000 feet down at Lava Falls. I had come to the forested North Rim, the canyon's "other side." It is a private region; only the wind jostled me as I gazed downward.

At first glance the canyon appeared eternal and changeless, but the longer I looked the more evidence of time I saw: the white scar on the Redwall, indicating a rock slide probably heard by no one; a curtain of lava which has erased the appearance of sedimentary rock; an old precipitous cattle trail winding to a spring.

I had changed too. I was leaving Grand Canyon with a feeling of protectiveness about the wild, fragile, irreplaceable environment I had seen. I had spent a month here, and I would return. After all, I must try to follow the timetable set by Powell: "It is a region more difficult to traverse than the Alps or the Himalayas, but if strength and courage are sufficient for the task, by a year's toil a concept of sublimity can be obtained never again to be equaled on the hither side of Paradise."

A relief map of southeastern Utah resembles bark on a tree, full of grooves and ridges where the country's greatest concentration of canyons gnarls the landscape. Canyonlands National Park, established in 1964, remains primitive backcountry. Just 12 miles of paved road exist in its 527 square miles. Only a jeep trail penetrates the Maze section, a cat's cradle of vertical-walled canyons on the park's western edge. It takes ropes or horses or the agility of a bighorn sheep to explore this particular realm, once known only to outlaws.

From a mesa in the Island in the Sky district, I took in the panorama of the park. The rough-hewn expanse of rocks, dusty red as though fired in a kiln, dates mostly from the Mesozoic Era, the time of the dinosaurs, when deserts and tidal plains took turns spreading over this region. The Colorado River cuts the major canyon here, meandering like an entrenched Mississippi through the arid plateau. Just out of sight is its confluence with the Green River, the Colorado's main tributary, and farther downstream is Cataract Canyon, a caldron of rapids which provides a major attraction for white-water rafters.

Beneath Canyonlands is a vast deposit of salt, remnant of a stagnant inland sea. Malleable to pressure, the salt has been squeezed upward in some places to warp the land, or has been dissolved by subterranean water to cause the collapse and fracture of overlying rock. The artistry of erosion has followed, creating a rock garden of arches, columns, towers, standing rocks, mesas, buttes, and spires.

To come upon the spires, known as the Needles, in the park's southern section is like approaching a megalopolis of rocks. These pinnacles of red and white banded sandstone, some reaching 300 feet in height, cluster like skyscrapers around open grassland.

"Yep, that's what we call the Manhattan skyline," said Slim Mabery, a retired park ranger who was guiding me on a five-day jeep tour through the Needles district. Jeeping is a sport to Slim. While I sweated during what seemed like a slow-motion rodeo ride, he grinned like a cowboy in town for the weekend as he eased the vehicle up the ledges on Elephant Hill. Some

Canyonlands National Park

spots are treacherous, as will testify the fellow who recently broke his pelvis and shoulder blade when his jeep overturned on the hill.

As we picked our way through the network of canyons, we saw repeated signs of civilization—prehistoric civilization. What at first looked like boulders strewn along ledges turned out, on closer scrutiny, to be dwellings of mud and stone, tucked under cliffs like swallows' nests. The houses belonged to Indians who lived and farmed in the canyons when the climate was wetter, from about A.D. 1075 to sometime in the 12th century. On nearby rocks the Indians' weathered drawings appear, sometimes showing a herd of deer, or geometric figures dressed like Hopi Kachina dolls, or merely a spatter of red and white handprints. Their purpose is unknown, but Slim offers a theory. "People ask me, were they recording some event in history, telling a story, or just doodling. The answer is yes. They were probably doing all three."

In the 1950's oil and uranium prospectors tracked all over Canyonlands. Half a century earlier cowboys ruled the land, grazing herds of cattle in the canyons. To examine one of their old camps, Slim and I left the jeep late one afternoon to hike to Lost Canyon. The way proved tougher and longer than we had anticipated, and it was sunset before we reached the camp, with its rotting corral, graffiti from the 1920's, and scattered debris: a Dutch oven lid, a chuck box.

We decided to chance a shortcut on the way back, but darkness found us high on a canyon ledge, not exactly sure of where we were. Nighthawks buzzed our faces as we sat on the rocks and waited for the moon to rise. When the pale light cleared the rim, we gingerly climbed up and down, up and down, sometimes on all fours, until midnight—when we reached the jeep. The Park Service had sent a search party after us. We felt sheepish; we also confessed to each other our touch of fright at having been in such a vulnerable position in such an unforgiving land. "Thou shalt be humble," we agreed, is a canyon's first law.

Black Canyon of the Gunnison

A turkey buzzard helped me appreciate Black Canyon. Looking down into the dark, precipitous gorge in western Colorado, I was having trouble with perspective: The canyon looked smaller than I knew it was. Janis Gifford, a ranger at Black Canyon of the Gunnison National Monument, had pointed out that I was standing on the edge of a sheer 2,000-foot drop, that the boulders below were the size of two-car garages, and that Painted Wall, a cliff across the way, has taken climbers a week to ascend. Nothing worked. Then Janis said, "Look at the buzzard." I peered down into space, expecting to see a big bird. Instead I saw a creature that within the depths looked as small as a starling. The canyon suddenly took on enormous dimensions.

Black Canyon wears three faces. At the upper end of its 53-mile length, firs, spruce, and pines climb down the walls, and the river is ponded into reservoirs by a trio of dams. The lower end, with its bare, colorful rocks and arid climate, is a desert by contrast. The most spectacular countenance, however, falls in the 12-mile middle section, named a national monument in 1933. It is a forbidding gash, split by a white-capped river, and arrayed with cliffs of granitic and metamorphic rocks. The gray walls date to Precambrian times, more than 1.5 billion years ago.

The canyon is a "fortunate accident," says Wallace Hansen of the U. S. Geological Survey in Denver. "We had a hard crystalline core, an uplift of the region, and a river with a steep gradient. With that combination, we ended up with a sheer, narrow canyon."

Feeling like a mere spectator on the rim, I climbed down into the canyon with Ranger Dennis Neilson. After a fairly easy two-hour descent, we reached the foaming Gunnison River, choked with boulders, bound by cliffs. "The river is unfloatable," Dennis said with a grin of experience. "To travel through here is a matter of swimming, wading, climbing, and busting through poison ivy higher than your head."

We spent a night and a day on the dusky bottom, fishing for rainbow and brown trout, and clambering over rocks. The climb back, up a loose-rock gully, was a character builder. Three-quarters of the way to the top I called for a rest and collapsed in the shade. With halfhearted curiosity, I asked Dennis what kind of bird was circling overhead. "That's a buzzard," he answered matter-of-factly. I needed no further impetus to pull myself up and make it to the rim. Once again in Black Canyon I found myself indebted to a turkey buzzard.

Little Cottonwood Canyon

A breath of cool breeze was awaiting me in Little Cottonwood Canyon. I was fleeing the hot, airless Jordan River valley, 20 miles to the west, where Salt Lake City spreads out like a griddle. I had to drive slowly. The road was crowded with others who wanted to climb the 11,000-foot mountains or luxuriate in the alpine meadows. Traffic is even worse in winter. Little Cottonwood, in the Wasatch National Forest, offers some fine powder skiing, and skiers pour into Snowbird, a large, bustling resort, or the quiet lodge town of Alta.

Geologists, of course, see more than good ski slopes. The canyon's broad, flat floor and its steep walls mean one thing to them: glaciation. "The canyon was a well-developed drainage before the glacier came through," noted Donald McMillan, Director of the Utah Geological and Mineral Survey. "But the glacier scoured the rocks and carved the canyon into its distinctive U-shape."

Above the glacial moraine at the mouth of the canyon rise dark granite peaks. It is here that Mormons quarried rock for their temple in Salt Lake City, and here that they blasted vaults to house genealogical records. At the other end of the nine-mile-long canyon are ski lifts, wooden lodges, access roads, and parking lots, as well as mine tailings from a century ago, when deposits of silver and lead were pulled from the ground.

Early on winter days, booms echo across the slopes as a 75-mm recoilless rifle hurls charges into the mountainsides. The explosives dissipate drifts before major avalanches occur. "This is probably the most severe avalanche area in a developed region of the country," says Jim Head, of the Forest Service. "In some months more than 400 snowslides are reported. Some are small; others could destroy a ski lift or a lodge." Each day, Forest Service personnel and their counterparts from the ski patrol survey the terrain, and if need be, stabilize the snow on the sides of the mountains.

One summer afternoon I hiked to a meadow, discovered a patch of snow, and started glissading down. I slipped and tumbled over and over and soon found myself on my back, staring up at an ocean-blue sky, smelling the pines, listening to a mountain stream. It made no sense to move. The spell of Little Cottonwood had taken hold.

Tinted chilly blue by a wintry dawn, Little Cottonwood Canyon in Utah shows the unmistakable U-shape of a glacier's work. During three distinct stages, ice moved through here, scraping smooth the granite walls. The last great ice sheet retreated 10,000 years ago.

DAVID R. STOECKLEIN, MOUNTAIN GRAPHICS

Canyonlands National Park

Carved layers of sandstone form a labyrinthine landscape in the Needles district of Canyonlands National Park, Utah. Anasazi Indians, the Ancient Ones, once lived and hunted here. One prehistoric

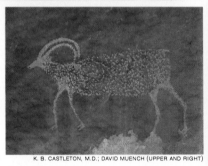

K. B. CASTLETON, M.D.; DAVID MUENCH (UPPER AND RIGHT)

artist etched in stone a spindly-legged desert bighorn (opposite, lower). Later Indians farmed the river bottoms, growing corn, beans, and squash. Granaries, built fortress-like in caves (opposite, upper), protected their crops. The Anasazi abandoned the area about 800 years ago, apparently after a prolonged drought.

Black Canyon of the Gunnison

Somber gray walls squeeze the Gunnison River and give the chasm its name: Black Canyon. Only 40 feet wide at the Narrows (right), the swift-moving Gunnison has cut a canyon 53 miles long and as deep as 2,400 feet in western Colorado. The sheer walls, carved from some of the oldest rock on earth, date from Precambrian times, 1.5 billion years ago. A gnarled juniper (left) writhes among boulders high above the river. The gray-green lichens (below) tinge rocks near the river. Lichens play a minor role in the formation of the canyon, producing an acid that gradually wears away the surface of the stone.

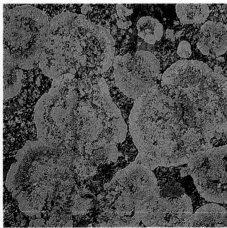

DAVID MUENCH

California and Nevada:

Silvered by a sudden autumn storm, evergreens edge the blunt masses of California's

Yosemite

Photographed by
National Geographic Photographer Joseph H. Bailey

Yosemite National Park—a 1,189-square-mile preserve in the Sierra Nevada.

GALEN ROWELL (BELOW)

86

Day packer pauses in Poopenaut
Valley near Hetch Hetchy, the rugged
canyon in Yosemite National Park
that naturalist John Muir called
"a wonderfully exact counterpart"
of Yosemite Valley. Muir battled to save
Hetch Hetchy, but lost: A dam on
the Tuolumne River just upstream from
this boggy meadow submerges eight miles
of his "grand landscape garden." Like
other roadless Yosemite valleys,
Poopenaut remains a seldom-visited
wilderness, a haven for wildlife.
The coyote population holds steady as
the wily creatures continue to find
sanctuary inside the park. Black
bears also thrive, having discovered in
hikers' packs a major source of food.
The bears usually avoid humans—
except to protect a cub, or when
they detect the smell of food.

From granite domes thousands of feet high, to the slight print of crustose lichens (right), Yosemite offers a wide range of spectacle. Its deep canyons derive from geological forces—folding, faulting, uplifting—that over the aeons produced steeper and steeper slopes; streams rushing down them eroded the sharp defiles. One such cascade near Tioga Road (far right) carries summer snowmelt. Glaciers also played a role in Yosemite's formation, cutting the canyons deeper and wider. Though the park's domes appear changeless, they erode continuously by exfoliation—the gradual peeling away of layers of rock.

"Photography is an appropriate use of Yosemite Valley," says Ansel Adams. Here the legendary photographer conducts a summer workshop in camera technique and "seeing"— as he has every year since 1955. His images of western grandeur, of the national parks, and especially of Yosemite Valley, have stirred millions of viewers. Adams relates his photographs to an exact time of day, acutely conscious of how light redefines and changes a scene, minute by minute. Cloud shadows mottle this late-afternoon view of Yosemite National Park from Glacier Point Road.

Flanked by famous heights—El Capitan (above) and Half Dome (below)—
Yosemite Valley has become a mecca for rock-climbers. Park visitors can
ascend Half Dome by grasping a system of cables, but the face
of "El Cap," 3,000 feet of sheer granite, challenges the world's best climbers.
Using fingers, toes, and ropes, an expert fly-walks toward the summit.

By
Michael W.
Robbins

I heard a scrabbling noise in the brush, and we stopped hiking. I turned to photographer Joe Bailey. "Could that be . . . "

". . . a bear." Joe finished my sentence and pointed. A fuzzy, half-grown cub was clinging to a nearby fallen log. Dappled by the afternoon sun, the young bear was a light cinnamon color. We stood still while it sized us up with sidelong glances. Then it jumped from the log and bounded into the undergrowth. "Where there are cubs there are mothers," Joe said. "Let's get out of here."

Suddenly "out of here" was a problem. We were at the bottom of a steep canyon in Yosemite National Park. Our car was only two miles away, but some 1,500 feet above us on the canyon rim. Ironically, we had ventured deep into the canyon to avoid the crowds and cars of Yosemite Valley. Now it seemed a place of solitude *and* danger. We hiked.

We had not gone far when I glanced back over Joe's shoulder. "Don't look now," I said, "but here's mama." A dark, full-grown sow bear had appeared, without a sound, behind us. For long moments we watched her and she watched us. She had a thick brown coat and tiny dark eyes. Sitting still, she did not seem threatening. It occurred to me that for the first time in my life I was within 20 yards of a wild animal capable of killing me.

Abruptly she lunged to one side and was gone. We climbed with renewed energy and tried to remember what we'd learned about black bears. Oddly, all we could recall was an ominously fragmented piece of advice we had received the day before. "When the bears get you. . . ." Neither of us could remember how the sentence ended. At every other switchback we stopped to catch our breath and to listen. No more bears, not that day.

To begin to explore the canyons of California and Nevada anywhere but in Yosemite National Park would be illogical. The Sierra Nevada and other mountain ranges in the two states are scored by many canyons, but Yosemite Valley is at once the most spectacular, the most accessible, and the most famous. We arrived during a spring of dramatic high water. The many recreational vehicles pulling into the parking lots heightened our concern about overcrowding, but the valley is only seven square miles of the 1,189-square-mile national park, and we could easily escape the crowds by venturing into the backcountry. With other first-time visitors, Joe and I were overwhelmed when we entered Yosemite Valley beneath Inspiration Point on the Wawona Road. We drove out of a tunnel, and there it was: the roaring Merced River 600 feet below, Cathedral Rocks and Bridalveil Fall, El Capitan, the green meadows and Half Dome.

Ansel Adams has devoted a lifetime to observing and photographing these and other landmarks, and few have seen more California canyons. We had caught a glimpse of him conducting his photography class in a glade, and later met him at the Ahwahnee Hotel. He is a burly man with a rolling walk and a ready humor. He strode the room in fur boots, recalling photographic views and vantage points. "Perhaps four o'clock is best for Lyell Canyon. Then you have sun in the valley. I got my pictures there about 3:30." He noted other views and areas, but always came back to Yosemite. He said, "Yosemite is the shrine."

That belief is widely shared. We found it verified when we met Charles and Alvine Adam from San Francisco. They were climbing ahead of us up

Western junipers cling to scant ledges above Yosemite's Tenaya Lake.
The short, sturdy trees thrive on rocky slopes, anchored by long, heavy roots
that help them withstand strong winds above the 7,000-foot level.

the steep horse trail toward Merced Lake. I noticed their serious and well-worn hiking boots and, resting at a switchback, struck up a conversation. "I've been coming here since 1921," Alvine told us. "Charlie and I have been making the trip together since the 1930's." On this hot afternoon they were day-hiking to Nevada Fall. The three-mile jaunt includes a 2,000-foot climb. Charlie is 80, Alvine 74. "We gave up backpacking five years ago," Alvine confided. The packs were becoming too heavy to carry at higher elevations. But nothing could prevent their annual summer pilgrimage.

Later, seated by a campfire in the valley, we met younger pilgrims who had traveled a great distance. One, a rock-climber so soft-spoken I could barely hear him above the crackle of the fire, had come from Australia. For three years he and his companions had planned this trip. They had come to climb El Capitan, the most challenging of Yosemite's cliffs. Now they had done it, had spent days concentrating with fingertips, toes, and ropes on the vertical granite. Yes, they would climb other mountains, probably in Colorado. But Yosemite had been their goal. "When you hear about climbing, you always hear about Yosemite," he said.

Yosemite National Park draws two and a half million visitors every year. So many come to the valley that Yosemite, like Yellowstone, is becoming a code word for an overused public facility. It could scarcely be otherwise. Concentrated in this one canyon along the Merced River, more than half a dozen peaks and domes rise from 2,000 to 5,000 feet above the meadows. Numerous waterfalls send streamers of pure, foaming thunder 300 to 2,400 feet down. Any one of these attractions would draw a crowd. To have them in a canyon where all can be seen from the central floor is what famed naturalist John Muir termed "a revelation in landscape affairs that enriches one's life forever."

During the Ice Age, snow fell faster than it could melt and changed to blue-ice glaciers. Rivers of ice formed in the heights of the Sierra Nevada and crept slowly down the valleys. The heavy ice carried so much abrasive rock that it cut the valley from a V-shape to the U-shape of today, and polished the bedrock to a smooth finish.

The canyons thus formed are havens for wildlife, including black bears, even though man's presence has altered this environment. I asked Len McKenzie, Chief Park Naturalist, about the changes. "There are probably three times as many bears in Yosemite now as there were before people came." They have found in backpacks and station wagons a major source of food despite park officials' efforts to educate the public in ways to keep food away from them. "The raccoons have displaced the ring-tailed cat in many park areas. They are more aggressive and have adapted better to getting food from human sources." The California bighorn and the mountain lion are now rarely seen here. Other animals have thrived. "Coyotes are holding their own. They're tough and resilient." Deer, squirrels, chipmunks, and some birds have grown adept at purloining food. One morning in camp I watched a bold Steller's jay swoop to peck energetically at a plastic package of nuts that I'd not yet opened; it had learned what such packages contain.

Len pointed out that human activity in the valley had also changed the plant life. "Cattle were grazed in Yosemite Valley even after the turn of the century. Native grasses were displaced by species introduced as stock feed. John Muir himself herded sheep in the area for one summer before he learned how they could ravage the land. He called them 'hoofed locusts.' "

Oddly, it was suppression of forest fires that worked the greatest

Merced River

Inspiration Point
5,391 feet

Wawona Road

Mount Watkins +
8,500 feet

Tenaya Creek

Tenaya Canyon

Yosemite Creek

CALIFORNIA

SIERRA

NEVADA

N

Yosemite +
Falls

Ahwahnee
□ Hotel

Mirror
Lake

Merced River

El Capitan
+ 7,569 feet

+ Half Dome
8,842 feet

Merced
Lake

YOSEMITE VALLEY

Bridalveil
Fall

Nevada
Fall

Cathedral
Rocks

Glacier Point
Road

Bridalveil Creek

*Carved by
water and
polished by
glaciers,
Yosemite
Valley draws
two and a half
million visitors
a year deep
into the
Sierra Nevada.*

changes in the valley flora. "Fire is an ecological necessity," Len said.
"Suppression of fire allows an understory of shade-tolerant trees to grow
up and create a fuel ladder that enables fire to reach the crowns of the
mature trees. It also allows the forest to encroach on the meadows. A hundred years ago there was about twice as much meadow acreage in the
valley as now." In what must strike some people as a betrayal of apple pie,
motherhood, and Smokey Bear, the National Park Service now *sets* some
forest fires—small "management" fires that clear the understory. "The
Indians did some burning. We are learning from fire research what they
knew from experience."

The Ahwahneechee Indians experienced Yosemite as a rich summer
abode, and in it they hunted, fished, and gathered acorns. In the late
1840's they began skirmishing with miners, and in 1850 they attacked trading posts in the Yosemite area that belonged to Maj. James Savage, a veteran of the Mexican War.

Savage led a quasi-military group, the "Mariposa Battalion," to sub-
due the Indians in their valley. Earlier white men had probably seen Yosemite Valley, but the Mariposa men were the first to venture in for a good
look. Their awestruck stories first made the public aware of Yosemite.
James Hutchings, who later published *California Magazine*, reached the
valley in 1855 with a horseback party. When their accounts—along with
sketches by Thomas Ayres—were published, they were picked up by newspapers and magazines all over America. Within a few years, small groups
were visiting the valley, and crude hotels were being built.

From the start, it was seen as a special place. Abraham Lincoln signed
a bill making it a state park in 1864, and it became a national park in 1906.
The Yosemite Valley Railroad reached the park entrance at El Portal in
1907. And in 1913, automobiles were officially permitted in the valley.

"Before shuttle buses were introduced, traffic was bad all over the valley," Ansel Adams told me. "Now I see people on foot, on bicycles, or maybe
in an occasional tour bus in the valley—it's a miraculous improvement."

The big crush came after 1970, we learned from Yosemite trail-
maintenance leader Jim Snyder. At the end of a long day of backpacking
above Merced Lake, Joe and I wolfed down more than our share of trail-
crew fare: roast beef, rice, gravy, salad, and coffee. We dined with
Snyder's men while he told me of weekends in the early 1970's when some
trails, such as the one to Nevada Fall, were trampled by 2,000 hikers a   97

day. "That's how you get messes like that trail down there used to be," Snyder said, referring to a stretch of the nearby Merced Lake Trail that for a time was a broad, mushy, lifeless bog. "In 1974, the park began restricting access to the wilderness areas, issuing permits. Now that same trail is limited to 60 or so backpackers a day."

A delicate balancing act is in progress: accommodating a large, restless, devoted public of all ages and interests, without letting them destroy the very places they have come to see. Today Yosemite Valley not only has cascades, domes, and deer; it also has gas stations, restaurants, camps, stores, a post office, a coin-operated laundry, a small emergency hospital, and parking lots. Ansel Adams concluded, "My first visit to the valley was in 1916, and I still think it's as beautiful now as it was then."

What remains miraculous is the ease with which one can escape all the development and be alone in the wilderness. "You can walk a few hundred yards and you're in the wild," Adams says. "I know you can do it," affirms historian Shirley Sargent, whose ridgetop ranch is in sight of the park. "I've done it. On a Fourth of July weekend I was able to get completely away from the crowds in the valley just by moving away from the roads."

Seeking a view into lower Tenaya Canyon, Joe and I set out one weekend morning from Mirror Lake. Within two miles we were alone in the place I found most memorable in all Yosemite: the Tenaya Creek cascades, below Mount Watkins.

Where the canyon narrows deep in a forest of ponderosa pine and Douglas fir, Tenaya Creek divided into several turbulent streams, fed by recent heavy rains. The main stream rushed over a rocky course with such energy that it entirely dissolved into lime-colored foam. We followed it upward, approaching a sound so thunderous that we could converse only by shouting. Ahead lay a jumble of rocks wet with permanent spray, surrounded by so many down-crashing cascades of blue and green and white water that we were nearly encircled by falls. All the streams flashed and danced with brilliance where they were touched by shifting beams of sunlight. It was as if we were faced with a frozen avalanche of granite boulders that was flooded by foaming white water. The sound was mountainous—solid thunder.

Except for one patch of warm sunlight, the forest was shaded, cool, and misty. That single zone of sunlight made it possible to stretch out and rest on the ground, which seemed to tremble with the force of falling water. The earth was strewn with pine needles, warped dry leaves, feathered and cracked pinecone fragments, gnarled twigs and bits of bark. It seemed a kind of brown salad.

The *scale* of Yosemite is deceptive. The volumes and drops are beyond what we are accustomed to measuring with a sweep of our eyes. Few of us are used to seeing, or feeling, so much solid rock so far above our heads. Nowhere in any other sierra valley is the sense of scale so dramatic as in the El Capitan Meadow. It is a flat field, traversed by a road. Behind the trees on the north side of the road is the base of "El Cap," as it's called. From that base the sinewy mass of granite rises directly 3,500 feet. The setback is often less than 200 feet; from one side of the face it is possible to drop a rock—or to fall—a clear 3,000 feet.

Most days during the season a string of parked cars lines the road, with drivers and passengers leaning on them, straining to see climbers on the face of El Cap. Climbers are not easy to spot. Even through binoculars they look as small as mites; were it not for their bright red or yellow jackets, they would remain invisible.

Often we stopped and looked for people on that enormous rock, and questions other than those of scale came to mind. Who could possibly be so careful, so strong, so nerveless? Once we stopped in the evening and saw the wink of a light about halfway up. Climbers were up there preparing to spend the night slung in nylon hammocks, anchored to the vertical rock. How could anyone do that? Who could sleep?

Seeking some avenue of understanding, and despite our admitted acrophobia, Joe and I signed up for the Yosemite Mountaineering School's course in basic rock-climbing. On a slow Monday morning our class consisted of just four people. Instructor Chris Falkenstein, with seven years of Yosemite climbing behind him, introduced us to the gear: nylon ropes, carabiners, and climbing shoes.

Chris spoke rapidly and expected us to get it right. One-inch nylon webbing made up the basic "swami" belt, to which rope and gear are fastened. Chris explained how to tie two simple knots, the water knot and the figure eight. We learned the basic movements on the rock: standing out from the rock face, using fingertips and toes, staying balanced, searching the rock for handholds and footholds.

Our first climb was easy, but our second was different. We were to begin with a combination push-up/chin-up onto a small ledge, then creep to our right on the faintest of cracks. Somehow we were to get over a bare hump of smooth stone the size of a Volkswagen Beetle, then onto a narrow ledge. From there we could reach a series of cracks that we could follow to the top. It was about 70 feet. "Just fingers and toes," Chris reminded us. "No fair hanging on the belay rope." He went first, climbing easily.

When my turn came, I was sorry I had worn shorts. I scraped both knees hoisting myself onto the first shelf. The big bulge of stone looked quite as smooth at close range as it had from below. It was difficult to stand out from the rock, and not hug it. I remembered that among the verbal signals climbers use was one we had joked about: "Falling!" Announcing your fate gives the person holding the belay rope a second to brace for the jolt. I hoped not to fall, but for long seconds I could find no combination of finger and toe traction that seemed likely to support my weight. Finally, I sprawled my way over the hump. The rest of the climb was simply exertion, a matter of lifting one's own weight. I began to admire real climbers, who had the stamina to ascend straight up for thousands of feet.

On the narrow ledge at the top my jubilation turned to fear when I saw that the descent would involve a rappelling exercise, directly off the edge. A straight shot, 70 feet down. It would be like stepping off the roof of my New York City apartment building. I had confidence in Chris and his ropes. The issue was whether I could get a steady enough grip on my nerves to step, backward, off the edge. I tied the belay line to my swami belt, then took up the rappelling rope I would hold as I fly-walked down the face of the rock. I slung it over my shoulder, right hand behind and left hand in front. A firm grip is important. If for any reason I let go, I would fall until the rope stopped me.

I backed toward the edge and stopped. I didn't really want to do this. Chris watched me and waited. "Look back to see where you're going," he said. It seemed unnecessary to mention that if I looked back over the edge I would not be going anywhere.

My hands got wet. Still I did not move. All my life I'd been afraid of something like this. Chris was patient.

"Look," I said. "I don't know if I can do this." He looked surprised.

"You can do it." What was the alternative? A helicopter rescue from this beginner's ledge?

Stiff with fright, I leaned against the rope and stepped off. It was springy but secure. I leaned back to grip the sheer rock with my hiking boots. Passing the rope very carefully through my hands, I walked down the cliff. I didn't look around or down, but concentrated on executing each step. It took many steps, but got easier as my confidence grew. When at last I reached horizontal rock, I stood still and light, and resisted an impulse to stoop and kiss the ground.

Joe came down with seeming ease, and we jabbered about the details of our descents. It had worked. Our admiration for the real rock-climbers, who spend days on cliffs dozens of times higher, acquired a new dimension. Afterward, when we saw tiny lights far above the rising dusk in El Capitan Meadow, we stopped to stare and stood shivering in the dark.

Roy Post stood in front of his cabin, smiling under his yellow cap, and recounted his two arduous winters as a ski-mounted mail carrier. "I packed the mail 'tween here and Quincy, on my back," he said. "I wore ten-foot skis and carried one pole. Used it as a brake. They paid me three dollars a day. In 1918 that was a good wage. I made the trip in about eight hours, three round trips a week."

Post packed the ten pounds of U. S. Mail over the 17 miles of the Quincy-La Porte wagon road at the upper end of the Middle Fork of the Feather River, in northern California. Then it was a stage road that wandered the ridges among gold-rush camps. Today it winds past abandoned mines and ghost towns, weedy townsites and miners' burying grounds, and in these last days of June the snow had melted just enough for us to get through in a four-wheel-drive wagon.

History is accessible here, and evidence of the gold boom days persists if you look for it. We stood talking in the shadow of bare hills scoured by hydraulic mining: Roy Post and I, naturalist Warren Grandall from the Plumas National Forest, and an old friend and hiking companion, Bob Goff. To Roy we mentioned our plan to explore Cleghorn Bar, a remote stretch of a Feather River canyon. "Oh boy," he said. "There's all kinds o' gold in there. My father had a claim there; we sunk a shaft." He went on dreamily, "I worked in lots of mines. . . ."

We bounced along rutted tracks through tall stands of ponderosa, thickets of gooseberry and chokecherry, chinquapin and manzanita. There were clumps of corn lily, and mule-ears with their yellow flowers and, now and again in a clearing edged in the shadows with old, dirty snow, patches of black-eyed Susans. At the bottom, we hiked upriver and began to find detritus of human use—all of it protected by the Antiquities Act of 1906. Some of the artifacts are a full century old. A level square of log chips and brick fragments, plus the half-shell of a gray enameled coffeepot, marked the site of a miner's riverside cabin. Across the river, a russet iron steam boiler with thick plates and pointed rivets recalled Civil War steam engines I'd seen in Vicksburg.

Farther upriver we found the rusted remains of a substantial mining operation, including, on its side, a boiler from a steam engine. I knelt to read the cast label on it: "Lidgerwood Mfg. Co., 96. Liberty Str., N.Y." I stood in surprise. That's only blocks from my Manhattan loft. It began to rain, and as we bucked up the slick, muddy trail we talked about the grit of the bygone miners. It seems to have known no bounds. We imagined them in bitter winter, riskily skidding machines down to the freezing river.

Middle
Fork
of the
Feather
River

We had ugly weather of another kind to contend with. The clouds opened at last, and we began hearing a sound dreaded by all in the forest: rolling thunder and the crack of lightning.

The storm hastened the end of the daylight, and it was nearly dark when we began traversing the rim of a deep tributary canyon. I noticed an orange glow, and watched it for some seconds before the obvious struck me. "Warren," I said. "There, to our left. Isn't that fire?" We got out binoculars. A lightning-struck tree flamed red-orange in the twilight and sent up twists of smoke, blue against the white fog. Warren hurried to his radio and reported it and another fire we spotted farther up the canyon. Later, we learned that these were among 28 confirmed fires reported that day. All were surveyed, but none developed into a serious blaze.

On my last day along the Feather I hiked to Bald Rock Dome, a granite knuckle jutting over the canyon rim. I settled to trace with my eyes the turns of chasms and folds of mountains, to watch the sunset shadow line rise up the far wall. I had done this in Yosemite and at Lookout Peak over Kings Canyon. My return hike would start at dusk and end in darkness. It was an acceptable bargain. The long view down a wild canyon, when you are the only human in sight, is worth some trouble.

Lamoille Canyon

At the head of Lamoille Canyon, at the 10,000-foot level, I found the perfect campsite, a room-size plateau with one wall of scrub pine and another of sloping rock. It was a balcony carpeted with pine needles, and beyond the rock wall the ground dropped away 200 feet to the snow-banked shore of Lamoille Lake. The lake was collared by a substantial snowpack, and the clear, deep water reflected the craggy edges of the bowl that is the very end of the 12-mile-long canyon in Nevada's rugged Ruby Mountains.

Few hikers passed—a threesome with bright red packs, three young couples who pitched tents on the lake shore—and I remembered what Al McElhiney of the Forest Service had said: "On any summer weekend, the Ruby Mountain scenic area has more beavers than people."

When you enter Lamoille Canyon between the steep mountain flanks, the canyon unfolds in a long crescent to your right. Lamoille Creek remains mostly hidden at the bottom in a green thicket of aspen, and the canyon sides are a sheer rise of 2,000 feet of bare, rugged rock, to spare, wild heights that wall out the rest of the world.

After several days of exploring, I noticed on the map that one of the hanging valleys, about 1,000 feet above the canyon floor, contained a small body of water—Island Lake. On a bright windy day I climbed to it.

The switchbacking trail led past an icy, downbound stream that was ideal for cooling feet, and along natural rock gardens of wild flowers. Island Lake proved a perfect round pond, as clear as any turquoise cove in the Caribbean. Its crowds of visible trout remained indifferent to my every lure and bait.

The half circle of snowy peaks above seemed temptingly close, so I climbed toward them. For 90 minutes I made good progress, but the ridge overhead looked no closer. From a broad, snow-covered ledge I could see how to finish the climb: another 300 feet up a snow-packed slope, then a 200-foot climb up a jumble of boulders and ledges. It was steep, but there appeared to be good handholds all the way, and I kicked footholds in the heavy, sticky snow. Below, Island Lake was a wind-rippled disk.

The rock was clean, dry, and firm, not brittle. The wind grew stronger, and each handhold called for careful decision. But there were no impassable stretches, and I pulled steadily upward. When I stood among the wind-dwarfed pines on top, a hundred-mile panorama of Nevada desert reached clear to the next mountain range, and the next, and the next. All of it was as rosy as the light of a fairy-tale illustration, with the near canyons in a cloud shadow. Patches of bright green in the farthest valleys melded up into the flat blues and purples of whole mountain ranges. A storm sweeping in from the west changed the light, blazing into the canyons below and dropping the distant ranges into misty grays.

It was one of those moments you know will be with you for the rest of your life, when you see something of the world as you've never seen it: canyons below, even mountains at your feet, the horizon as far off as it's ever been, and a roiling, stormy sky just overhead. Reasons to seek canyons, to climb mountains.

Cathedral Canyon

As a rule, no water flows through Nevada's Cathedral Canyon. I stood on the ruins of a dirt road washed out by a series of flash floods, and the loudest sound was the pounding of blood in my own ears. No moving water, scant wind, few trees to rustle. Certainly no human sounds intruded. Cathedral Canyon is tucked into the White Pine Mountains, 60 miles from Eureka and Ely, the nearest towns. A mile-long classic canyon, it's an experience in near-perfect silence.

It wasn't always quiet. A century before, it had echoed to the sounds of hooves, creaking leather, and the conversation of high hopes and bitter losses. The road had been built from silver mines at Treasure Hill out across the mountains and flats to Tonopah. Supplies for an 1870 population of thousands of miners were carried up from a rail link, and silver was freighted out.

Scattered across the mountains are the hollow shells of ghost-town banks, hotels, saloons, cabins, and livery stables. Hamilton, Treasure City, Eberhardt, Monte Cristo, all are little more than names. Kicking among the sage that carpets these sites confirms that more than bread and tools were freighted in. I found bottoms of wine bottles, carved buttons, a woman's shoe, bits of English china.

The canyon entrance is as distinct as a doorway, and I camped before it on the washed-out road. Sundown brought a darkness as complete as the silence. On a 360-degree horizon of broken hills there was no light but starlight. The snap of the fire was the lone sound, and its orange flicker, reflected on a close circle of sagebrush, conjured scenes from a hundred Western films.

In the morning, great horned owls as somber as gargoyles observed my stroll along the canyon's washed-out floor. High on the craggy walls, layers of rock had eroded, forming arches and battlements. Swallows swept from one wall to the other and, high above, hawks hovered. Near the rim, the dark red rock had worn through in window-like openings, and lines of hard white quartz looked like petroglyphs. At shoulder height along the wall I found a den, musky with hair and droppings. Across from it was a mine shaft no larger than a well, its dry cribbing holding back the gravel tailings. How long had it taken to excavate? I wondered. And what hopes had brought men to this Cathedral? As with all canyons, its allure is deeper than its walls, broader than its mere expanse.

Hot summer silence fills Cathedral Canyon, an arid defile in Nevada's White Pine Mountains. Cathedral once echoed with sounds of the fevered traffic of silver miners. Now, soaring owls and hawks share its narrow air.

MICHAEL W. ROBBINS

ALL BY MICHAEL W. ROBBINS

Contending with stubborn granite, the Middle Fork of the Feather River twists out of the northern Sierra Nevada. Along the ledges life finds a niche: A western fence lizard basks—vigilantly—and mountain pride spills color from sheltering fissures.

Lamoille Canyon

Aglow with autumn light, Lamoille Canyon
harbors a fertile oasis along the only paved
road into Nevada's Ruby Mountains. Its

entrance concealed by slopes and a glacial moraine, the canyon surprises visitors with
a scenic, 12-mile crescent of aspen groves, beaver ponds, and meadows along
Lamoille Creek. Trout lakes and desert vistas reward hikers on its rugged trails.

Canyons of the Tarahumaras

live in Mexico's rugged Sierra Madre Occidental.   *Photographed by Tor Eigeland*

Train-borne tourists encounter Indian artisans at El Divisadero, a whistle-stop overlooking the barrancas—the canyons of the Tarahumaras. Before the Chihuahua-Pacific Railroad pushed through the mountains in 1961, the Tarahumaras found few markets for their baskets, pottery, or violins. Now the train makes a 15-minute stop, and travelers disembark to view the gorges and bargain with the Indians. Vendor Victoria Torres (right) may find a buyer for the bear-grass baskets she weaves. *Following pages:* Serrated ridges of the sierra rise and fall for thousands of square miles, discouraging all but the hardiest hikers—and the indefatigable Tarahumaras.

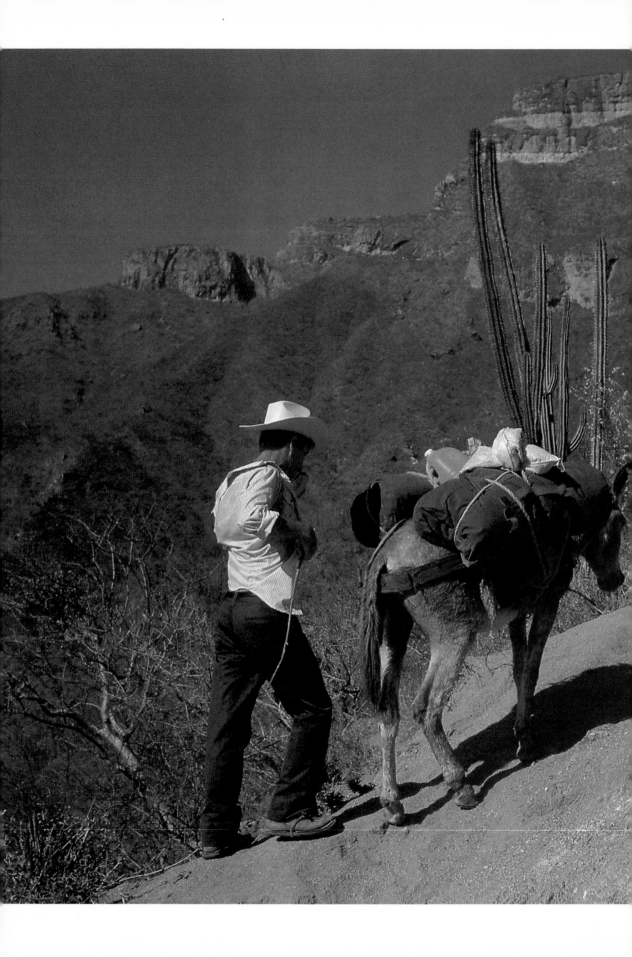

Pack-laden donkeys
lead the way up
a narrow, dusty trail,
urged on by guide
Rosario Díaz. The
everyday footwear of
the Tarahumaras,
sandals cut from old
tires, protects his feet.
From an airstrip at
Creel, the author made a
long, sweeping journey
through the sierra.
Rosario guided him for
this leg, from Batopilas
to Urique, a trek of
about 20 miles that
required two and a half
days of travel. Once
trails such as this
offered the only way into
the barrancas. In the
1960's, the railroad
began bringing some
visitors to the area, and
now roads have begun
linking scattered towns.
More and more tourists
arrive, drawn by the
spectacular scenery.

116

"Halfway to Heaven," Tarahumara farmers utilize every acre on a mesa high above the Barranca del Cobre—Copper Canyon. Many Tarahumaras cultivate such plots atop the mountains in summer, and move to warmer canyon floors in winter. Oxen provide the power for tilling, dragging wooden plows like those introduced by the Spanish 300 years ago. Farmers may plant squash or beans, but corn remains the staple: fermented for *tesgüino*, mild beer for ritual and social use, or drunk as *pinole*, a watery gruel. Graciana Martínez (lower left) grinds roasted corn on a *metate*, a concave stone of a shape in use in southwestern North America for thousands of years. She will mix the corn flour with water to make pinole.

Refreshment for man and beast comes with a bath in the Urique River, as Rosario Díaz washes down a mule. This reward for Díaz, his animals, and the author marked the end of the arduous journey from Batopilas (right), where the 100° F. temperature emptied the streets at midday, to Urique on the river. Both towns began as Spanish mining settlements around 1700. For generations, animals and men packed silver out of the mountains. The mines closed when the veins played out, and remoteness preserved the towns. About Batopilas, one resident said, "We used to feel we were in paradise." Now, new roads will bring changes, perhaps renewed mining.

DAVID HISER (BELOW)

Living up to their name—*rarámuri*, or "foot runners"—
Tarahumara men compete in a kickball race that may go
on for 200 miles. In such major races two teams, each from
a district of *ranchos*—or settlements—race many laps
over a two- to twelve-mile cross-country course. Team
members take turns kicking aloft a wooden ball. All day,
friends and supporters come along just for the run, and
carry torches all night to light the kickers' way. In this
double exposure the torches show as streaks of flame.
Magic, dietary taboos, heavy betting, and the inevitable
corn beer surround an important race. Shorter and more
informal contests can be run in a single afternoon.
Tarahumara hunters once caught deer by running after
them nonstop until the animals dropped from exhaustion.

By
Tor Eigeland

Slammed by downdrafts, kicked by updrafts, shaken, buffeted, and rattled, I was attempting to take photographs out of the open window of a small plane. I threw a quick glance at the graying pilot, Captain Sergio Molinar Vásquez, looking for reassurance. A slight yawn cracked his expressionless face. Reassuring. The best pilot in the sierra, it was said. As Molinar himself had told me: "Here you're either great or you're six feet under."

This bumpy reminder of the rough, unforgiving nature of the abruptly spectacular canyon country below us was not really necessary. It was the sixth time in 17 years that my career as a photojournalist had brought me to the Barranca del Cobre—or Copper Canyon—in the region known locally as the Sierra Tarahumara. I have always felt an affinity for the world's hot places, and welcomed the opportunity to return to the canyons of this part of Mexico. Though remote, the Sierra Tarahumara is one of the places in the world where I am assured of a warm welcome by old friends.

Located in the southwestern corner of the State of Chihuahua in Mexico, the sierra is home to some 50,000 Tarahumara Indians and an uncertain number of mestizos, or people of mixed blood. My return now, in June 1978, was the first leg of a journey to the sun-baked canyons of Mexico and Texas. From Chihuahua I would turn east to the Big Bend country of Texas, then north to its Panhandle.

Gray smoke shrouded the bluish-black volcanic canyon walls, the green pine and oak forests on top, and the scarce vegetation farther down. The rainy season, which should have begun by now, had failed, leaving the sierra tinder dry. We counted no fewer than 11 forest fires in the Barranca del Cobre area alone.

Speeding through the smoky skies in relative comfort, I found it impossible not to reflect on the fact that in less than an hour we had flown over terrain that might require torturous weeks, if not months, to cover on foot. I have made seven-mile mule trips here that turned into three-day ordeals. The land of the Tarahumaras is among the roughest in the world. Many peaks rise more than 8,000 feet, and the largest rivers have carved barrancas as deep as 6,000 feet. Erosion has sculpted bizarre buttes and pinnacles. The walls of the canyons are cave-riddled tiers of yellow, pink, brown, tan, and white strata. In places there are canyons within canyons, small barrancas stairstepping downward into a craggy, mile-deep chasm.

We landed in unpredictable crosswinds at Creel, traditional gateway to the sierra. Here an old friend, Father Luis Verplancken, a Jesuit missionary, waited to welcome me as he had on several previous occasions. Over dinner that evening, I asked Father Verplancken whether there had been any great changes in the sierra. "Well," he said, "the people are still growing their corn, beans, squash, and some fruit and vegetables. They still have their goats and a few head of cattle here and there. But there are now roads through the sierra. This has forced big changes, not necessarily for the better. For the Indians, these roads have provided a lot of jobs. But visitors are bound to be attracted by the beauty of the land and its remoteness, and are certain to exploit the area and the Indians. It has to happen. All we can do is try to educate the Tarahumaras and help them retain their own culture—which gives them their unity."

Father Verplancken pointed out that the newly built roads and the

Authority among the Tarahumaras resides in elected gobernadores such as Patricio Ramírez. Serving without pay or privileges for an indefinite term, each gobernador resolves disputes and delivers sermons and judgments. 123

Chihuahua-Pacific Railroad that stops at El Divisadero have made it possible to promote tourism; already a few hotels have been built above the canyons. This has been a boon to the Tarahumaras, said Father Verplancken. "It used to be nearly impossible to obtain any kind of Indian craft. Only baskets were readily available. Lately the Tarahumaras have been selling a lot of pottery, violins, carvings, and dolls. It has been a good source of income, a welcome change for the Indians."

A frightening new aspect of life in the sierra, I found, was the traffic in drugs. Many local people, tempted by big money, have begun growing opium poppies as well as marijuana. The traffickers are dangerous, and it is most inadvisable for tourists to stumble into the wrong areas.

A more welcome change for me was the wide dirt road that led out of Creel. The old road had been narrow and often impassable. Early one morning I set out from Creel with driver Oscar Loya, heading for Batopilas. From there, I planned to make my way by foot, by truck, and by train to El Divisadero. Yawning chasms flank this little settlement, perched at the confluence of the major arms of the Barranca del Cobre.

The sun was just coming up as we drove through the cool pine and oak forest of the high sierra, at about 7,500 feet. Though very dry, the ponderosa pines were giving off their fine, refreshing scent. After some 35 miles we descended to a point where a new bridge was being built across the Urique River. The stream is still young at this point, and the jade-green water, with its pools and banks, looked idyllic and inviting. I suggested a swim, but changed my mind when construction workers told us that they didn't bathe at this point for fear of being attacked by a colony of otters.

We climbed the south side of the canyon to another plateau at about 7,500 feet. Stretches of pine forest, mixed with oak and reddish-brown madrona trees, alternated with plots of land planted mostly in corn, which was beginning to wilt because of the delayed rainy season. In the much warmer climate down in the barrancas, the farmers hold off their planting until it is actually raining.

The people, mostly Tarahumaras, were staying out of the summer heat. Most of them live in *ranchos*—settlements that may consist of just one family or as many as 20. We saw a few men and women sitting in the shade of their log houses and others at the mouths of caves. Some little boys and girls kept the hungry cattle and goats out of the cornfields by tossing rocks at them. Just past the rancho of Quírare, we came to an obligatory halt on the rim of the canyon of the Batopilas River. Some people stop to pray here. Others just look, or do both. The road narrows to little more than a rocky trail poised above an abyss. The view is dizzying—but magnificent. I was reminded of the reaction attributed to Father Salvatierra, probably the first white man to see the Barranca del Cobre, in the late 1600's. "At sight of the precipice, such was my terror that I . . . did not precisely dismount, but let myself fall off on the side opposite the precipice, sweating and trembling all over with fright." Far below lies the Batopilas River and the tiny town of La Bufa, a mining settlement that is now almost a ghost town. Oscar Loya brought the car down to La Bufa without incident, no doubt disappointing the buzzards circling above us. As the sun had climbed, so had the temperature. Here below, the heat was suffocating, ranging around 100° F.

Until recently, La Bufa was the end of the road for automobile traffic. From here it was a long day's hike on mule trails along the river to the old

Urique
River

Batopilas River

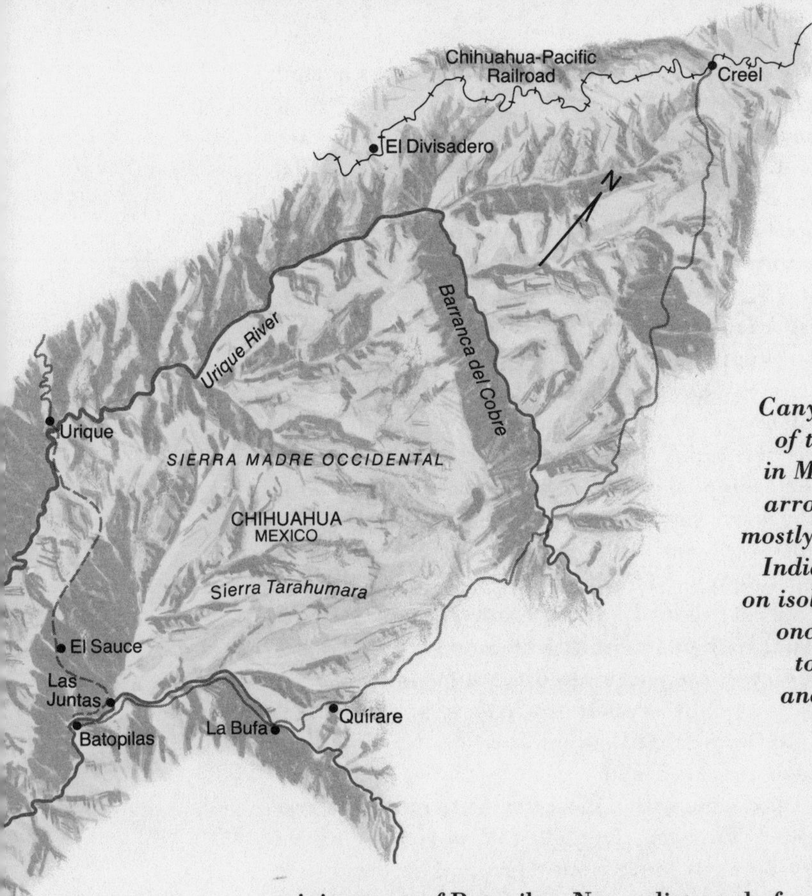

Chihuahua-Pacific
Railroad

Creel

El Divisadero

N

Urique River

Barranca del Cobre

Urique

SIERRA MADRE OCCIDENTAL

CHIHUAHUA
MEXICO

Sierra Tarahumara

El Sauce

Las
Juntas

Batopilas

La Bufa

Quírare

*Canyons rend the mountains
of the State of Chihuahua
in Mexico. Sun blisters deep
arroyos here, and rain falls
mostly in summer. Tarahumara
Indians eke out an existence
on isolated ranchos. Old towns,
once mining centers, hope
to boom again as more
and more tourists arrive.*

mining town of Batopilas. Now a dirt road of sorts reaches Batopilas, and we made the drive in something under an hour.

Luck awaited me right at the edge of Batopilas, a town of some 600 people stretching in a thin line along the banks of the river. Stopping at the first house to ask directions, I recognized Rosario Díaz sitting on a mule. Five years earlier he had guided me to the site of an isolated Tarahumara Easter celebration.

I definitely needed pack animals and a guide to cross the mountains from Batopilas to Barranca del Cobre, and here was Rosario again, grinning broadly, asking: "Need a good mule skinner, señor?"

The road has brought no traffic jams to Batopilas. Only a mail truck passed by while we were there. A few cows wandered in the street, and a little troop of pigs busily sniffed for tidbits.

The first time I saw Batopilas it reminded me of a South Pacific island. The vegetation is lush. Most of the people are dark-skinned. Children swim in the wide river, and women gather on the banks to chat or wash clothes. Trees planted through the centuries mercifully shade the streets. Flowers bloom everywhere.

Founded in 1708, Batopilas soon boomed, and for something like 200 years silver and other metals were mined here. Isolation and dwindling reserves finally made production too expensive, killing large-scale mining. Everything had to be transported on the backs of animals or people. But isolation has also preserved an Old World flavor. In spite of the lack of amenities and the scorching summer heat, I like Batopilas.

How long can this unspoiled quality last, I wondered, now that it's possible to arrive by car. A young woman, Señora María del Pilar López, said, "We used to feel we were in paradise. We were alone here with our beautiful mountains and canyons, and we almost never saw outsiders. It is

125

gone now with the road." Later Manuel Alcaraz Garner, in whose home I was staying, spoke enthusiastically of the possibilities that the new road brought: "It may be economical to operate the mines again, and there are plans to promote tourism and convert an old hacienda on the river into a luxury hotel."

But local color still abounds. That evening I had dinner at a "restaurant" that was really just the living room of an old adobe house. Very straight-backed chairs and a big table with a plastic tablecloth accommodated the paying customers. A faded picture of the dark Virgin of Guadalupe hung on one wall. Piglets kept bounding through the doorway onto the dirt floor, but a young waitress just as quickly booted them out again. I was greeted courteously, but nobody asked me what I wanted to eat. There was only one menu: tortillas, refried beans, a hot chili sauce, and fried eggs. The menu was the same for each of the six meals I ate there. To drink, there was either warm water, warm Coke, or warm canned juices.

About four in the morning I was awakened briefly by unearthly screaming and thrashing outside my window. Later I asked Don Manuel what had happened. "Our parrot has died," he said sadly. "He was such a good companion for everyone. He had a fight with an animal . . . we don't know what." The town character, the parrot used to visit every house in Batopilas. Whenever there was a procession, such as a wedding or a funeral, the parrot would circle overhead, squawking loudly. Batopilas was genuinely mourning the death of a friend.

That evening Rosario Díaz came with a mule for me to ride, and two donkeys for packing our gear. We spent the night at Las Juntas, where Rosario lived. We could get an early start from there and accomplish a good part of the climb before the heat of midday. Urique is only 20 or so miles west of Batopilas, but we planned on at least a two-day hike.

The temperature hovered around 80° F. when we left the river at Las Juntas at seven in the morning. As we climbed, so did the sun. Step after step we struggled up rocky trails, across crumbly shale and slippery rock faces, dodging spiny plants. Acacia, mesquite, organ cactus, agave, and prickly pear grew alongside the tracks.

Dripping with sweat after a climb of some 2,000 feet, we arrived at a little rancho called El Sauce. A wiry, graying woman of about 50 was standing by her wooden gate. We introduced ourselves, and the woman, María de Morales, smiled and waved us to a shaded porch where she brought a large earthenware jug of cool spring water. Rosario and I quickly finished it, and Doña María filled another. Into this one she squeezed sour oranges, then added a little sugar. It was the most delicious drink I had ever tasted.

Then, once again, up, up, up. I alternated between climbing on foot and riding the mule. At about 7,000 feet we reached the shade of a pine forest, and the soothing breeze was as refreshing as the orange drink.

Two days later we were within an hour's walk of the Urique River. We had long since descended from the cooler, pine-forested regions, and it was now hot again, although still early in the morning. The few poor ranchos we passed were growing mangoes and other tropical fruit.

As we neared the river, Rosario and I became more animated, and so did the donkeys and the mule. They picked up speed, and my mule even began to run. Though heavily loaded, the donkeys also started trotting, for the first time on the trip. The river was within sight, and the air carried a heavy, intoxicating scent of tropical vegetation. Mad with excitement, the animals raced. Reaching the bank, I leaped from the mule, threw off the saddle, stripped, and jumped into the pale-green water. The mule and Rosario followed, but the more decorous donkeys remained on the bank.

Easily fording the shallow river several times, we reached the town of Urique in half a day. Stunned by the humid noon heat, its 800 inhabitants had vanished inside their houses. We rode to the center of town without seeing a single human being. Only two mules and a few pigs stirred. Somehow I had expected Urique to be unchanged since my last visit in 1962. The big trees were still benevolently shading the streets and the gaily colored tin-roofed houses, but to my surprise the town had grown considerably.

Ramón Figueroa, an old acquaintance, told me later that since November 1975, when a road of sorts had entered Urique, many little houses had been built, not by new settlers, but by former residents who had returned to Urique. With no industry, only subsistence-level farming, some cattle ranching, a bit of trade, and the odd water-driven mill in the river producing a few grains of gold, the Urique area is at the moment dirt poor. Its people live largely on hope.

The story of Urique is virtually identical to that of Batopilas. Founded by mining interests around 1690, the town produced silver, gold, lead, copper, and zinc. But the isolation made modern mining unprofitable. Nonetheless, mining still obsesses the people of these canyons. Nearly every visitor is approached by a dreamy-eyed character who knows of a mine full of high-grade ore. "I only need a little capital to work it," he says.

With the new roads, mining may bring improved economic conditions, and increased tourism may help. The community leaders in Urique speak wistfully of investors to provide capital for restaurants, hotels, and other services. The possibilities are certainly good. It is sad to contemplate, though, what large-scale mining and tourism would do to these tiny, self-contained old communities. I mused on the future of Urique while sitting on the riverbank. Two lovely, raven-haired girls of 11 or 12 came up to me and said, "Would you like some coffee, señor?" They had seen me sitting there, and had run home to fetch coffee and cookies for me.

The Urique River lost its tranquil aspect that afternoon. Towering thunderheads built up above the canyons and exploded with tremendous fury. Giant bolts of lightning flashed across the skies, and thunder crashed and reverberated between the canyon walls. The accompanying deluge lasted for hours. This signaled the end of my stay in Urique. "If you want to get out by road," Don Ramón said that night, "you had better make it soon. This afternoon there were several rockslides. We'll open the road again tomorrow, but after that, who knows? It may be closed until the end of the rainy season."

As the next afternoon's downpour commenced, to the accompaniment of more thunder and lightning, we pulled out of Urique in an old truck. Children and adults, for some reason, applauded us. Before starting the precipitous 6,000-foot climb, on a narrow road that is one switchback after another, the driver crossed himself. So did an Indian woman who rode with us, two babies asleep on her lap. I tried to roll up the window on my side to keep out the rain, but there was no window. There were no windshield wipers either, and I could barely make out the road in the driving rain. We ground our way up about 4,000 feet, crawled over a few minor rockslides, then came to a halt. The driver had just asked me to remove a boulder from the road when the truck began rolling backward. We had no brakes, either! Expertly, the driver turned the wheel and the rear of the truck slammed into solid mountainside. "We almost went back to Urique," he said.

A few more hours' muddy drive and a short train trip brought me to El Divisadero. Here the chasms of the Barranca del Cobre meet. Weathered mountains and peaks, rocky plateaus, and dark gorges stretch to the horizon in three directions. Clouds send their shadows sliding across the terrain, creating moving patches of darkness and light. The whole magnificent panorama looks impossibly hostile, as though a giant had crumpled the landscape like newspapers and flung them randomly about.

El Divisadero had undergone a transformation since my last visit in 1962. There was no station here then, but passenger trains would make a quick stop so tourists could run down to the edge of the canyon and have a quick look at the magnificent abyss below. This time I stepped off the train onto a platform and into a swarm of souvenir vendors. There were shops and stalls run by mestizos, and Tarahumara women sat on the ground, their voluminous, multilayered skirts, their baskets, pottery, and wood carvings spread out around them.

Just below the station a new hotel hung on the edge of a cliff overlooking the Barranca del Cobre. The best of two worlds, El Divisadero provided a good bed, a shower, and proper meals, along with unspoiled wilderness a few hundred yards away.

The Tarahumaras appear little-affected by the tourism. For a few minutes every day at the train station, goods and money change hands. But there is no time for anything else, and most of the Tarahumara vendors speak little beyond the most basic Spanish, if that. The outside influence on the Tarahumaras has come not from tourists, but from the resident white and mestizo population, from government officials and teachers, and from the Jesuit missionaries. Except for a few thousand who have resisted, the Tarahumaras have converted to Catholicism or, perhaps more correctly, have added Catholicism to their own beliefs. Some of the men wear trousers and sombreros instead of breechcloths and headbands, and many have learned a few phrases of Spanish. Oddly, a number of mestizos have picked up some Tarahumara habits. Many brew the traditional *tesgüino* of the Tarahumaras, and they sometimes drink *pinole*—a mixture of roasted, ground corn and water. Often the handshake of the mestizo ranchero is close to that of the Tarahumara: Outstretched hands touch horizontally without clasping. And in the Barranca del Cobre I have seen both mestizos and Indians participate in the traditional Tarahumara kickball races.

That night I sat on the edge of the Barranca del Cobre enjoying a magnificent sky and a multitude of scents triggered by the afternoon rains. Strange night sounds rose from the black chasm below. And looking up at the stars I remembered a legend a Tarahumara had told me: The stars are Indian girls, it says, very beautiful, who were shot by arrows into the sky. The girls cling to the arrows, and adorn the heavens.

Santa Elena Canyon

D riving toward Big Bend National Park, 275 miles east of El Divisadero, on the Texas-Mexico border, I picked up a radio station that called itself "The Voice of the Last Frontier." I chuckled at the presumption, but later, after spending some time in the park, I made a mental apology to the announcer. For the Big Bend is indeed one of the few places I have found in the U. S. where it is still possible to spend several days without seeing another human being.

Some 700,000 acres of mountains, desert, and canyons, Big Bend National Park once sheltered Indians, conquistadors, and bandits. In 1916 one band of outlaws caused an international incident by crossing the river and raiding two towns in Texas.

Santa Elena Canyon forms part of the splendid park. Between 75 and 130 million years ago, the Big Bend area was covered by a sea. As ocean creatures died and sank to the bottom, they gradually formed thick limestone layers. Later, forces within the earth thrust this limestone upward, cracking it and shaping it into mountains. Over thousands of years, sand and gravel in the flowing water of the Rio Grande eroded the limestone to form the canyon. Santa Elena is more than 1,500 feet deep and 17 miles long; it is less than a quarter of a mile wide at its top, and in some places narrows at the bottom to barely 50 feet.

Just before dawn one day I stood at the mouth of the Santa Elena watching the sun come up. The Rio Grande, about 50 feet wide at this point, spilled out of the canyon. As the sun rose, the coloring of the canyon walls, reflected in the slow-moving waters, gradually changed from faint blacks, browns, and warm, rusty reds, to grayish-white. Before long the sun moved on and left the canyon in shade. I strolled half a mile down the Santa Elena Canyon Trail, which begins here, and felt dwarfed by the high, massive, sheer walls. But more than anything, I remember the sounds. Like a well-orchestrated symphony played with restraint, songs of birds and insects flowed from the lower canyon, then vibrated and echoed up and out into the sky. Looking up, following the sounds, I could see black vultures soaring up and down the canyon walls. It looked almost as if they were dancing to the songs.

Mariscal Canyon

Some 30 miles downstream from Santa Elena, at the very bottom of the Rio Grande's Big Bend, is a shorter but slightly deeper chasm, Mariscal Canyon. As in Santa Elena, mesquite, willow, tamarisk, tree tobacco, and reeds grow in a strip of green along the water's edge. White-winged doves, canyon wrens, and cardinals flit among the branches. Visitors occasionally spot rare and elusive peregrine falcons soaring overhead.

The Mariscal is ideal for boaters with little experience—a category that includes me—so I signed up for a six-hour float through the canyon with rafter Rod Ponton. From the time we launched our craft into a crack in the mountains, we entered a strange world where the shadows are nearly black, where the scenery is spectacular, and where, according to Rod, "there are no serious hazards. It's possible to take a spill or two," he told me, "but the only really difficult spot is appropriately named the 'tight squeeze'—the channel narrows to barely ten feet there."

We floated along, enjoying the scenery and the solitude, rowing a little from time to time. At one point the rock face rose in sheer 1,600-foot walls beside us. Some of the colossal boulders, nearly white, looked remarkably like abstract sculptures.

Rod pointed to a wooden platform perched high up among sheltering rocks. "This country was once so isolated that a hermit made it his home. He lived up there back in the late 1960's," he said. "One day, the story goes, he spotted a life jacket floating in the river and ran down to retrieve it. Buckled into it was the body of a young woman who had drowned. The hermit was so shocked and disturbed that he packed up and moved out."

Midway through our trip we came to another appropriately named spot: the "break." The canyon walls here are about a quarter of a mile from the river. A dry creek bed disappears into Mexico, and trails wander into the mountains on both sides. Indian petroglyphs on a boulder testify   129 to the great age of the trail. Here too, on the Mexican shore, is the abandoned camp of a party of wax makers. Rod explained: "The candelilla is covered with a wax that helps it retain moisture. When the plant is boiled, and sulphuric acid is added to the water, the wax floats to the surface. Packaging and selling it was a big industry during World War I."

The Rio Grande slips out of Mariscal Canyon as abruptly as it enters, and meanders through the desert. We hadn't seen another person—smuggler, ranger, tourist, hermit, or border patrolman—all day long.

Palo Duro Canyon

From Big Bend I journeyed some 450 miles north to a canyon in the heart of the flat plains of the Texas Panhandle. For mile after mile, the grassy countryside rolled beneath my car, until it seemed there would be no end to it. Then suddenly, with an abruptness that took my breath away, I found myself at the edge of a canyon. Some 750 feet below me the Prairie Dog Town Fork of the Red River—a small and gentle stream—undulated the length of the canyon. Over the last million years it and its tributaries have performed a heroic task, carving a gash 45 miles long and as much as three miles wide in the flat, windswept countryside of the Texas Panhandle.

I drove down into the canyon and was in another world. Stands of juniper and cottonwood shaded the banks of the little stream, where scattered mesquite and hackberry grew. Multicolored mudstone slopes and sandstone cliffs rose beside me. Concrete fords cross the stream here and there, but sometimes, after heavy rains, the Prairie Dog Town Fork shows its power, and flash floods roar through the canyon. Animals live in this natural shelter, and for ages man has come to hunt them. Stone weapons found in the canyon indicate that people were here 12,000 years ago. Indians made their lodgepoles and arrows from tough juniper brush, and the juniper gave the canyon its name: Palo Duro means "hard wood" in Spanish. Later, ranchers used this same wood to make fence posts.

The first European to see the Palo Duro was probably the Spaniard Francisco Vásquez de Coronado, who set out with an army in 1540 in search of a legendary land of silver and gold, but found only Indian settlements. During the expedition, the Spaniards camped in a ravine that may have been Palo Duro Canyon.

In the 18th and 19th centuries, traders and buffalo hunters often stayed in the canyon, and in 1874 one of the last of the large-scale battles of the Red River Indian War took place here, when U. S. Army Col. Ranald S. Mackenzie defeated a confederation of Comanche, Kiowa, and Cheyenne Indians, driving them from this part of the plains. Just two years later, rancher Charles Goodnight realized that the canyon would make a natural corral, and drove 1,600 head of cattle into it.

More than 16,000 acres have been designated Palo Duro Canyon State Park, where visitors come to hike, ride, camp, and picnic. Standing at an overlook one evening as the shifting sunset performed its old magic, tinting the rocks with a palette of rusts, purples, pinks, and grays, I marveled at an apparent paradox: In areas often thought of as dry—the Panhandle and the Big Bend of Texas, the tortured terrain of Chihuahua—water has used the epochs to sculpt monuments from stone and earth, to stab with walls of beauty the very land.

Abrupt gash in a flat plain, Palo Duro Canyon stretches 45 miles across the Texas Panhandle. The Prairie Dog Town Fork of the Red River carved the canyon. Normally a trickle, the stream occasionally erupts in flash floods.

Santa Elena Canyon

Watchful hummingbird nests in a tamarisk tree in Santa Elena Canyon. Located along the Texas-Mexico border, the 17-mile-long canyon straddles the Rio Grande in a dark cut 1,500 feet deep; in places it narrows to just 50 feet. At its mouth, the canyon affords space for a trail to emerge from sheer, echoing walls into the sunlight. The flowering shrub, feather plume—a *Dalea*—and the brilliant orange flowers of the caltrop, or Mexican poppy, brighten the canyon.

STEPHEN J. KRASEMANN (ABOVE AND OPPOSITE)

Mariscal Canyon

Rafters drift through Mariscal Canyon, like Santa Elena a part of Big Bend National Park; mesquite and reeds fringe the river. The desert, habitat of wild flowers and cactus plants (below), spreads a carpet of blossoms in times of rain. Moisture prompts a startling transformation in the resurrection plant (lower right). When dry, it remains as tightly closed as a fist, but a rainy night brings a "resurrection" of unfurled fronds. A voluble black-throated sparrow (lower left) greets the desert morning.

ED COOPER PHOTO; STEPHEN J. KRASEMANN (LOWER LEFT); DAVID MUENCH (LOWER RIGHT)

Time and a river combine to create a row of columns along the Upper Iowa River. This 70-mile-long gorge, 450 feet deep in places, surprises visitors to "flat" Iowa.

The Midwest:

Gorge of the Upper Iowa

Photographed by Matt Bradley

Afternoon sun slants across Marcellus Broghamer's 200-acre dairy farm near Decorah, Iowa. In the valley beyond flows the Upper Iowa River. Farms occupy 80 percent of the land along its banks. The last glacier to touch this part of Iowa retreated a million years ago; the hilly terrain remains as the legacy of weathering and stream erosion. To minimize loss of soil, Broghamer plants his corn and hay in alternate strips, harvesting each at a different time of the year. At far right, Tearza and Lisa, two of Broghamer's five children, shoot baskets against a windmill that once drew water from a well 325 feet deep. In 1939 electricity made the windmill obsolete. Today, some 20 million people live within 250 miles of the Upper Iowa, and cows and canoeists alike share its waters.

Tenacious shrubs and trees cling to dolomite cliffs
along the Upper Iowa; the vegetation's spreading
roots contribute to the crumbling of the wall. Here
author Ed Welles and geologist Jean Prior canoe
the gentle stream. Laid down in ancient seas that
once covered the Midwest, the rock yields fossils
nearly half a billion years old. Signs of more
recent life along the river: Cliff swallows peek
from nests built of mud. "The birds were in
almost constant motion," says the photographer,
"wheeling and darting overhead." At lower right,
anemones sprout from a limestone ledge. Indians
incised the petroglyph, probably of a buffalo, at
least 400 years ago. Such weathered carvings
appear in nine shelters in the soft sandstone along
the Upper Iowa.

By
Edward O.
Welles, Jr.

The wind rose, faceting the sunlight on the river. Overhead, clouds billowed and sailed across the sky. We moved away from the bank and into the urgings of the current, then stopped paddling and let it take us down through the land, full and green with the season. It was the first of June, a diamond-bright morning.

Dick Baker, a geology professor at the University of Iowa, and Matt Bradley, a free-lance photographer from Little Rock, Arkansas, floated ahead in one canoe. I rode with Jean Prior, a geologist with the Iowa Geological Survey. Jean, I would learn over the next two days and 20 miles of river, knew the Iowa landscape like the back of her paddle, and could speak of a time when rivers slept beneath blankets of ice as thick as the hills. As we drifted, she explained that the northeastern corner of the state, through which the Upper Iowa River runs, had escaped all but the earliest of the ice sheets.

Thus, for the last million years or so, the terrain here had been shaped by the subtler forces of weather and water—often with unsubtle results. We rounded a bend, and the bank, which until then had been close and wooded, suddenly rose a hundred feet in a wall of dolomite. The looming stone amplified our voices, and we lowered them to savor the deep silence. "Dolomite is very resistant," said Jean. When the Upper Iowa, on its 135-mile journey from southeastern Minnesota to the Mississippi, periodically wanders up against it, the two lock horns and the rock yields grudgingly. Jean pointed to vertical gashes in the wall. "Those are joints. They're natural fractures, or partings, that extend through the rock. The freezing and thawing of percolating water, and tree roots growing down through the cracks, enlarge them. This, combined with the force of the river striking the joints at an angle, causes the rock to fall away in blocks, rather than wear down along smooth contours."

We pulled alongside an enormous pile of rocks that had crashed into the river. The dolomite lay in jumbled slabs, each large enough to raft us down the river. But time had loaded them with a ballast of fossils. We climbed onto one, and Jean pointed to a honeycomb design in the stone as intricate as its name. "That's a *Receptaculites*, a primitive marine organism; you could also call it an 'index' fossil. If you came here and didn't exactly know where you were, geologically speaking, it would tell you."

These fossils began forming nearly half a billion years ago when shallow seas washed over much of the continent. The life forms were pressed between layers of sediment like flowers between the pages of a book. For millions of years the seas flowed and ebbed over Iowa before retreating and leaving layers of sedimentary rock more than 4,500 feet thick in places.

We moved on from there, down through the afternoon. The river wound past farms and through woodlands, now and again flowing beside bluffs of dolomite rising from the shore like a row of massive Doric columns. The sun lowered, and as the wind died a stillness settled on the river. A great blue heron took off from the shallows and sailed away downstream. A muskrat slipped into the water.

A little farther on we rounded a bend and surprised a huge great horned owl from a shadow high in the stone. It seemed to fill the sky. Just as it crossed the river and disappeared into the trees, a second owl

Simple span from the 19th century bridges the Upper Iowa near Bluffton. Immigrants from Norway, Germany, and the British Isles settled here in the 1850's. Such wrought-iron bridges moved their grain to market; now the spans carry campers to grassy parks along the river.   143

followed. But this one fluttered and lost altitude, finally crash-landing in a young willow on the near shore. The little tree wobbled under the weight as the bird clung to it, riding its momentum. We paddled over to stare. The current swung our canoes; paddles resounded against aluminum gunwales.

The owl was a fledgling. In another week it would have made it across the river. It fixed big, yellow marble eyes on us, then lost its balance and dropped a couple of rungs in the tree. To halt its decline the bird unfurled its wings into a feathery rust and white mosaic—there in the green tree beside the white rock. We moved away quietly.

The gorge of the Upper Iowa is a popular scenic and recreational area, and the river itself was found eligible for inclusion in the National Wild and Scenic Rivers System. It is a beguiling stream for canoeists who don't need the thrills white water offers. Small riffles alternate with long, quiet pools. Near its headwaters, the river is a meandering prairie stream that crosses and recrosses the Iowa-Minnesota border. Limestone bluffs and outcroppings begin to appear, the valley deepens, and the river's gradient increases to more than seven feet a mile. The river grows wider—from 30 to 70 feet.

Between Kendallville and Bluffton, palisades, bluffs, chimney rocks, and forested slopes rise along the river. Occasionally, farms reach down to the banks, and cows and horses stand swishing their tails and watching canoeists pass by. The river runs through Decorah, but earthen levees, built to protect the town from floods, block most of it from view. Near the Mississippi, the terrain changes again. A broad, flat-bottomed valley opens up, flanked by steep bluffs as much as 450 feet high. The valley seems very spacious here, after the narrow reaches upstream.

This river caught the eye of settlers in the mid-19th century. They considered its swiftness and built a number of mills along it. Some were sawmills, others woolen mills, but most ground the wheat, corn, and buckwheat that was planted on adjacent lands. Little wheat is grown here now, but agriculture remains the predominant way of life along the Upper Iowa. Corn and hay are among the big crops. That is what Marcellus Broghamer grows on his 200-acre dairy farm half a mile from the river near Decorah. "I like farming," he told me. "It's the seasons. You go through winter and can't wait for spring. Then you work hard outside in summer and fall, and you're ready for winter again and doing chores inside." Broghamer plants his hilly land in alternating strips of corn and hay. Sowed and harvested at different times, each crop holds the soil until the other has a chance to take root. The steepness of the terrain along the Upper Iowa makes the soil very susceptible to erosion.

Perhaps no place on earth offers a better lesson in how canyons are formed than a steeply rolling midwestern farm, where a furrow can quickly become a gully, which in turn can grow and deepen until it has destroyed an entire pasture. Jim and Wendy Stevens live in an old log house in a narrow valley that presumably was once just such a gully. A spring runs down through their valley, grows into a swimming hole near the house, and continues half a mile to the Upper Iowa. Jim and Wendy draw water from the spring for their garden, which yields more than 20 kinds of vegetables.

I visited the Stevens place late one afternoon. As we walked out to look at the bees they raise, Jim told me that midsummer is the time of year when "the main honey flow is on." It felt that way then. Angled sunlight filled the small valley, enriching the green of the grass underfoot.

Jim and Wendy also have an old lime kiln on their property. Now

Angling southeastward, then turning to the northeast at Decorah, the Upper Iowa wanders 135 miles and empties into the Mississippi River.

laden with vegetation, it hasn't been fired since the turn of the century. It burned limestone, yielding lime which in turn was sold to a paper mill. The mill was destroyed by fire in 1905, and the fortunes of the kiln cooled.

The landscape here was once studded with such historical artifacts. On the morning of our second day on the river, we put in at one—an old iron bridge. It and others like it were built across the Upper Iowa during the late 19th and early 20th centuries, supplanting less durable wooden structures. A web of wrought iron, with a wooden roadbed, this bridge looked ready for 21st-century traffic.

People came to the Upper Iowa long before they had perfected the art of building bridges. The first of them arrived about 11,000 years ago, drawn by game. The forested hills and valleys of northeastern Iowa sheltered more wildlife than did the adjacent prairie.

We were in search of man the hunter that day, taking out late in the afternoon at the confluence of Bear Creek and the Upper Iowa. That was where we met Clark Mallam, who teaches anthropology at Luther College in Decorah, and a few of his students. Mallam, now in his thirties, grew up in southeastern Nebraska, the son of a tenant farmer. Moving around with his family brought him and his restless curiosity into contact with a good portion of the Nebraska landscape, and with a lot of Indian artifacts that no one could tell him much about. Mallam persisted. The quest led him to anthropology and to Iowa.

When Mallam arrived here from the flat country he had known, he couldn't believe what he was seeing. "The hills were like mountains to me; I climbed every one I could find." His rambles over the Iowa countryside, as in Nebraska, brought him face to face with a number of Indian remains. And we were now headed for one of them.

We walked back from the creek toward the wooded slopes of the valley through a field of sprouting corn, and Mallam spoke of the people who were here before us and who left their marks—however faint they might now be. To the east along the Mississippi River Valley and reaching north into central Wisconsin lived the peoples of the Effigy Mound culture, semi-nomadic hunters and gatherers so named because of their tradition of building earthen mounds in the shapes of animals. Their culture lasted roughly from A.D. 700 to 1200. More concentrated where we now walked was the Oneota culture. It overlapped Effigy Mound in time and place, sharing the bounty of the Mississippi, and dating from about A.D. 800. "They were 'dual subsistent,'" said Mallam. "They were hunters and gatherers, but they also raised crops."

145

The trail led into the forest and up a slope. Fifty feet up the trail we came to a sandstone outcropping. Carved into it was a series of symbols. Some looked mathematical; others were animal images: deer, turtles, buffalo. Indians, probably the Oneotas, had carved these figures more than 400 years ago. The soft light of dusk lent scant definition to the badly weathered carvings. But as I ran my hand over the rock and felt its graininess, I remembered similar subtle pleasures: morning sunlight firing an old iron bridge, an owl flying from an impressive wall of rock.

"Anthropologists are at odds in interpreting petroglyphs like these," said Mallam. "Some believe they represent different social groupings; others think the rock shelters where they're found could have been ceremonial centers. It's impossible to determine their exact function. Perhaps their real significance for us is their beauty, and the relationship to nature they symbolize. I like them for that."

And so did I.

In 1964, parts of the Current River and one of its tributaries, the Jacks Fork River, were designated the Ozark National Scenic Riverways. Once they weren't very scenic. The piney hills these two Missouri rivers run through were heavily logged between the 1880's and the 1930's. Deer and wild turkey were largely replaced by cattle and hogs, which roamed the unfenced land. I recently canoed 30 miles on the Jacks Fork beneath hillsides that, again covered with trees, have restaked their claim to wildness. The river flowed clear, gliding over refractive chert as it moved beneath dolomite and sandstone bluffs. The rocks, austere grays and tans, rose out of hills rampant with vegetation.

The summer heat which had caused the hills to burgeon had also becalmed the river. March thaws and April rains were spent. The Jacks Fork was low, living off a reserve of springs which underlies this land. There are hundreds of major springs in the Ozarks. By August they are providing the rivers with as much as 75 percent of their flow. The springs nurse the streams to the new year, when they resurge with snowmelt and rainfall.

But now it was summer in Missouri. I rented a canoe, packed some food and a sleeping bag into it, and set off alone. If the heat and the solitude weren't enough to slow me down, the river would. As I floated the Jacks Fork, the easy movement of the canoe through a rapid often ended abruptly with the harsh protest of aluminum on chert. Out of the canoe I clambered, through the current I sloshed, dragging the recalcitrant canoe in search of deeper places.

The chert, a brittle stone, was layered in the bluffs overhead, and it had come away with the blocks of dolomite chiseled free by erosion. Into the river went the stone. The dolomite dissolved, but the resistant chert remained—to be mulled over and rearranged by the current.

Small-scale agriculture is the mainstay in the Ozarks. Farmers here raise mostly grass and hay for their cattle and hogs. While the place demands a flintiness of the people, it seems to expose few hard edges. In the Ozarks I met people who, like the river, had plenty of time. Their lives seemed unchanneled and relaxed. A few of the stores are general, selling everything from ice cream to overalls, as ceiling fans whir overhead. In a cafe, a remark to a stranger about the weather may lead to a life story. The woman from whom I rented my canoe also ran a grocery store. Her husband, a seasonal employee with the National Park Service, carpentered during the winter. In summer they helped their parents put up hay. Resourceful people with plenty of time.

Jacks Fork River

During my three and a half days on the Jacks Fork, the river, by its pace, insisted I not be a stranger. I stopped often to swim, to discourage the heat. I saw details—snakes sunning on rocks, the light filtering softly through the trees—that I might have missed on a more urgent trip. I stopped early each evening, camping on gravel bars, to allow time to watch the sun set through the gorge; to see the full spectrum of shadows created by sunlight angling against rock. Then, with darkness descending and fog brewing out over the river, I would scratch out a place in the gravel and fall asleep, happy at being in a lonely place.

Little Missouri Trench

As we reached the top of a butte in North Dakota and had our first glimpse of what lay beyond, Neil Korsmo said, "There's been a fire up here." The place hardly suggested that. It was grassland, deep and green after the spring rains. It moved with the wind to a line of buttes rimming the sky. I believed Neil, though. During his three summers as a ranger at Theodore Roosevelt National Park nearby he had fought six minor prairie fires. "Have you ever seen a juniper tree burn?" he asked. "It catches fire around the base, its shaggy bark ignites, then suddenly it bursts into flame. It's kind of unnerving." It was a scattering of skeletal juniper trees which told Neil that there had been a fire here. They stood like crosses in the rejuvenated field.

Below us moved the Little Missouri River. Before the Ice Age, it had flowed north toward Hudson Bay. The glaciers sliding down out of Canada bent it around to the east, aiming it ultimately for the Gulf of Mexico. This increased the river's gradient and thus its power. It gouged hundreds of square miles of Badlands from the plains. It was along a stretch of the river called the Little Missouri Trench that we now stood. We had a day of canoeing behind us, and it felt good to get up into the land. From the river, the buttes had seemed painted by an artist whose respect for the straight line had matched his love of color. The result was a banded landscape, as startling as a rainbow. Strata of rock—gray bentonite, black lignite coal, red scoria, tan sandstone—filled the canvas. Cottonwood trees along the river and the blue sky overhead framed it.

The Badlands, where rainfall averages a mere 14 or 15 inches a year, come by their name naturally. By mid-July the prairie grasses are scorched the color of ripe wheat, and the river is too low even for a canoe. Long winters and dry summers have historically made ranching a chancy business. The Sioux battled the U. S. Cavalry here, skirmishing with them on adjacent plains, then leading them into the rugged Badlands, where the soldiers and Indians fought each other to a standoff.

The last big sorting out in these parts began during the Dust Bowl days of the 1930's. The wind blew the topsoil away, and grasshoppers harvested much of the wheat. Cattlemen were reduced to feeding their herds tumbleweed. Many people left—the land and life foreclosing on them.

One who stayed was Marjorie Solberg. She works at the general store in Grassy Butte, where I met her. She came to Grassy Butte in 1918 as a three-year-old, her parents having moved up from South Dakota to homestead a quarter section—160 acres. "They were looking for something better," she told me. They found an unfenced range, a land that in good years could fatten a lot of cattle. But Marjorie's home would not see electricity until 1952, the telephone until 1959.

Marjorie married a cowman in 1933. They scraped through the drought, ranching on rented acreage. In 1943 they bought a ranch and an additional 320 acres for ten dollars an acre. They reared three sons who

147

are now grown and gone. Marjorie's husband died in 1972. She leases the land to a neighbor, but continues to spend her summers in the red cedar ranch house, built in 1920 of logs hauled from the Missouri River.

She told me all this as we walked through her pasture one evening. The grass was emerald in the dying light. "I've never seen it better than this year," she said. "I suppose an old lady like me has no business staying on alone out here. But my roots are so deep."

The St. Croix River rises cold and dark in the silent forests of northern Wisconsin. One of our National Wild and Scenic Rivers, it flows west, then south, for 165 miles before emptying into the Mississippi River. On its way, it passes through the Dalles of the St. Croix, a narrow and rockbound gorge, like sand flowing through the center of an hourglass. The river, roiling with glacial debris, carved the gorge into an ancient bed of basalt at the close of the Ice Age. The Dalles are less than a mile long, a moment of geological emphasis.

Walking through the gorge early one morning, I rounded a shoulder in the riverbank and came upon a hollow strewn with fallen boulders. I began working my way back up through this rock garden, jumping from one boulder to the next. A large one teetered under my weight. An adjacent boulder shifted, then a third. In a long moment filled with the groaning of the rock, I wondered if I had precipitated a minor geological event.

Such an experience would have been all in a day's work a hundred years ago. Then loggers worked these banks, and the land was a deep forest of white pine. The mammoth trees were cut and, with high water in spring, floated to sawmills below the Dalles. But often, in tumbling through the bottleneck of the gorge, they would jam. One logjam in 1886 required a steamboat, 24 horses, 200 men, and dynamite to break it up. The logs were stuck for six weeks. One of the sawmills was 17 miles downriver at Marine on St. Croix, Minnesota. Built in 1839, it was the first commercial mill along the river. The previous year, David Hone and Lewis Judd had ascended the Mississippi and then the St. Croix. Lured north by reports of land opening up for settlement, they found a site for a sawmill where a creek rushed into the river.

Built in three months, the mill sawed its first log in August 1839. By 1855 it was milling two million board feet of lumber a year. But it was operating by the sketchy rules of the frontier. A depression followed a bank panic in 1873. A 57-day logjam occurred in 1883. The year after that, low water marooned the logs upriver, and a tornado blew the smokestack off the mill. Starved for capital and credit, the operation went bankrupt.

What is left of the mill lies at the bottom of a ravine. One August afternoon I scrambled down the hillside through a tangle of vegetation, looking for it. Mosquitoes homed in on me. I found the waterwheel, rotted and lying on its side, and parts of the stone foundation. The creek ran clear and cold—the way Hone and Judd must have found it. I sat there awhile, listening to the creek, contemplating the patterns of vine and moss described on the old rock. The vegetation and the crumbling mill seemed to combine, creating a memorial of sorts to man and also to nature.

Chill blue waters of the St. Croix River mark the Minnesota-Wisconsin border. Blocks of basalt, cut and polished by the river, form the Dalles of the St. Croix. French-Canadian voyageurs named the Dalles, using their word for gorge. Surging with glacial meltwater 10,000 years ago, the river carved this mile-long gash in a layer of solidified lava.

148

PERRY RIDDLE

ALL BY PERRY RIDDLE

Jacks Fork River

Dolomite bluffs rise above the verdant banks of the Jacks Fork River in southern
Missouri. Dolomite and sandstone line the upper half of this 45-mile-long stream. At
left, a crayfish on the riverbed and a partly submerged frog accent the water's clarity.
Springs nourish the Jacks Fork and honeycomb the land. Geologists, spelunkers, and
landowners discover scores of water-carved caves in the Ozarks every year.

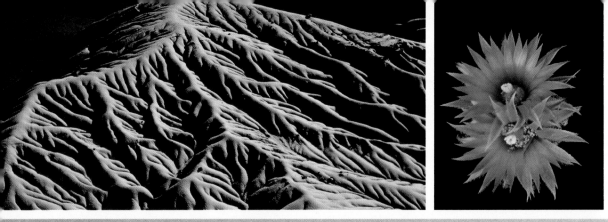

Little Missouri Trench

Seamed with lignite coal, Badlands buttes flank the Little Missouri River in North Dakota. Details from this worn land: a sandstone frieze (right) and weathered siltstone (far left). A ball cactus adds a perennial burst of life.

ALL BY ENTHEOS

Canada and Alaska:

Twin chutes of roaring water thunder around a limestone pinnacle at Virginia Falls,

Canyons of the Nahanni

a 294-foot cascade in Canada's Nahanni National Park.

Mist of an August dawn swirls over Virginia Falls. Downstream, the walls of three forested canyons rise thousands of feet alongside the South Nahanni River. Hot springs and ice caves near the river represent opposing forces: expiring volcanism and cold winters. Pressure from Canadian conservationists resulted in the formation of the park in 1972. They opposed the building of a power plant at the falls, and the mining of the area's copper, zinc, and coal. Wedged into the southwestern corner of the Northwest Territories, Nahanni remains accessible only by boat or bush plane. Fewer than 300 people visited the park in 1978. Those who did found the wildlife abundant and varied. The willow ptarmigan (left), which turns entirely white in winter, ranges across the North American Arctic.

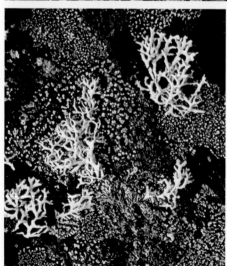

From thaw in May to frost in September, diverse and hardy plants sustain wildlife along the Nahanni. Birds eat crowberries (above, left), and black bears graze on bearberries (above, right). Lichens flourish throughout the park, providing food for goats and caribou.

Silt-laden South Nahanni hairpins through The Gate in Third
Canyon, below Virginia Falls. The walls here rise 700 feet from the

river. The Nahanni gorges began forming nearly two million years ago
as the river sliced into uplifting layers of limestone and dolomite.

Songs from guide Dirk Van Wijk's harmonica and the echoing shriek of a gull (right) pierce the silence of Nahanni. Early in this century, reports of gold lured prospectors up the South Nahanni, but the rumored discovery of headless corpses fostered legends of murderous Indians and savage mountain men. Adventurer R. M. Patterson helped dispel the tales when he investigated the area in the 1920's and wrote of its splendor: "Never in my wildest dreams had I hoped to see anything like this."

STEPHEN J. KRASEMANN (ABOVE); BO RADER

Backed by a mountainside spiked with aspen and spruce (right), boaters on a South Nahanni beach prepare to camp. River runners must find campsites on high ground, because the water can rise a foot an hour during heavy rains. In 1978 UNESCO officials added Nahanni National Park to their World Heritage List. They noted the area's unusual geological features—especially the hot springs, seldom found this far north.

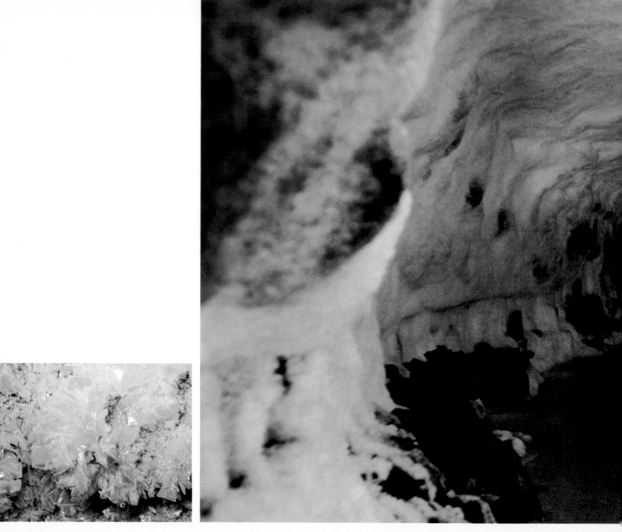

Icy tunnel threads Grotte Valerie, most
extensive of the 150 caves that honeycomb
the limestone walls above the South
Nahanni. First explored in 1971, Grotte
Valerie has 6,300 feet of mapped
passageways. At left, the author's party
prepares to enter the west portal,
the cave's only easily accessible entrance.
Inside, freezing temperatures create
glistening caverns and banks of sparkling
ice crystals (above, left). In another
corridor (right), bones tell a poignant tale
of Dall sheep that evidently became
trapped at the base of a slippery ledge.
The remains of more than 100
animals have been found, some as
much as 2,000 years old.

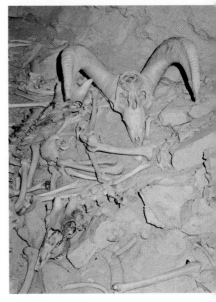

Like the steps of an ancient amphitheater, terraces of Rabbitkettle Hotsprings rise in a spruce forest near the park's northwestern boundary. Water bubbling from the spring (below, center) cascades down the steps, leaving deposits of tufa that over the last 10,000 years have formed a mound 90 feet high. Frost and erosion by the Rabbitkettle River nibble at the mound. Visitors (lower left) tread carefully; their footsteps can shatter the fragile tufa. A willow sprig (lower right) survives in solitude on the surface.

165

Alerted by an unfamiliar scent, a moose
cow and her calf pause in a lowland
thicket. Indians of the Nahanni
once used the small pouch of hairy skin
under a moose's neck in a ritual:
After a successful hunt, they hung the
pouch—or bell—in a tree near the
scene of the kill to invoke good fortune.
A mountain goat (right) bounds
across a near-vertical wall. At left, a red
fox dozes in a meadow. Trappers
once hunted the sleek animals for their
pelts. Now rarely seen, foxes
search the canyons for field mice,
cottontails, insects, and berries.

By
George Joseph
Tanber

Around the turn of the century, rumors began filtering out: The skeletons of two murdered prospectors had been found in one of the valleys along Canada's South Nahanni River. Their remains were scattered, and it looked as though someone had tried to destroy the corpses. The heads were missing. Within a few years, other men disappeared in the Nahanni wilderness, and the legend began to build. Popular journalists published articles about the "River of Mystery" and the "Valley of Vanishing Men." More reports came in, of people who had ventured into these canyons in the southwestern corner of the Northwest Territories and had never been seen again, of burned-out cabins, of massacres. Even the place-names hint of tragedy and evil: Funeral Range, Deadmen Valley, Headless Creek, Devil's Kitchen, Broken Skull River.

So it was with some trepidation that I made arrangements along with photographer Bo Rader to explore these canyons—especially since I was a neophyte canoeist, and the rapids along the Nahanni are dangerous and seldom run in canoes. We would also venture into other canyons in Canada and Alaska—northern, pine-cloaked gashes where loons fish from gold-flecked waters, and mined-out silver veins tell of long-gone bonanzas.

Derek Brown and Dirk Van Wijk, two young canoeists from Ottawa, agreed to guide us on a three-week, 350-mile river trip through the Nahanni country. Few people ever see Nahanni National Park, a 1,840-square-mile reserve along the South Nahanni. It can be reached conveniently only by plane or boat, so visits to the park are expensive.

It was midsummer, the best time for a river trip in Canada, when we set off for the headwaters of the Nahanni. With our canoes lashed to the pontoons of a twin-engine plane, we circled 7,466-foot Mount Wilson and came to rest on a small lake in a swampy area at the base of the mountain.

"The first 50 miles of the river offer some of the best white-water canoeing in Canada," said Derek as we maneuvered our canoes through narrow, grass-lined channels. The river was a cool lime green. For three days it lived up to Derek's claim. We challenged rapid after rapid, charging down a stream that was in near-constant turmoil.

Gradually the river changed as the miles hurried by, becoming wider and tamer. Silt-laden tributaries, some fed by melting glaciers, rushed in from both sides, turning the stream turbid and cold. Loons occasionally darted overhead, startling us with their piercing cries, and bald eagles glided in to perch in tall spruce trees and watch as we drifted past.

It took five days of canoeing to reach Rabbitkettle Hotsprings, one of Nahanni's most astonishing scenic wonders, on the northwestern boundary of the park. We camped on a sandbar in a light rain and a cloud of mosquitoes, and in the morning paddled across the river. The trail to the springs led through a thick forest of spruce and poplar, crossed ankle-deep muskeg that squished underfoot, and zigzagged through heavy brush.

We found David Wright, an ebullient, red-haired Irishman, camped near Rabbitkettle. A graduate student from the University of Alberta, David was doing research on the springs. A mound 90 feet high and 200 feet across rose before us, its terrace of tufa stairsteps a blend of off-white, apricot, and iron red. "It's been forming here for something like 10,000 years," David said as we climbed toward the spring at its center. "Water from deep in the earth bubbles out and trickles down the flanks, gradually

Slow-moving porcupine clings to a branch near the South Nahanni. More than 40 species of mammals inhabit the park, and flocks of migratory birds make rest stops on the numerous lakes and ponds.

STEPHEN J. KRASEMANN

depositing calcium and other minerals as it goes. Careful where you walk. It's very fragile." Chunks of tufa that had broken off lay scattered around us. "Because of the harsh winters, it's unusual to find so spectacular a hot spring this far north," David told us.

Evening found us back on the river. Golden rays from a setting sun drenched the valley as we paddled toward our next goal, Virginia Falls, 80 miles downstream. On the shore, a porcupine waddled into the brush; overhead, arctic terns squawked as they dove for fish. We passed a forest ravaged by fire, its charred trees still standing, like black tombstones.

The next day toward dusk we heard a faint rumble—muffled thunder. Virginia Falls was within sound, if not sight. We camped that night a mile above the falls and in the morning portaged around them. Through an opening in the dense forest I saw the river pick up speed as glittering white water tumbled toward the precipice. We circled around and down to the base of the cataract, 294 feet high, divided in midstream by a jutting limestone pillar. A cloud of mist, shimmering against the morning sun, hung high above. "Not surprisingly, Virginia Falls is the most popular attraction in the park," Superintendent Eric Hiscock had told me earlier. "Most visitors are canoeists, who normally spend two or three days there, but some people who visit the falls hire an outfitter and come upstream by jet boat."

No story of the Nahanni would be complete without mention of a legendary prospector named Albert Faille. At the confluence of the Flat and Nahanni rivers, we entered the country he once had roamed. "One of Faille's old cabins is still standing a few miles up the Flat," Derek said as we beached our canoes north of the junction. Born in New Salem, Pennsylvania, Faille spent 30 years prospecting and trapping in the country around the Flat River and above the falls. Though he never struck it rich, he managed to make a living selling pelts. English adventurer R. M. Patterson wrote in 1954: "The Nahanni has probably never seen a finer canoeman, and to watch Faille search out the weak spot in a riffle and plant his canoe's nose exactly there . . . is like watching a fine swordsman seeking for an opening, feeling out an adversary." Just before he died in 1973, Faille was asked what sort of park he thought Nahanni should be. "Let it stay wild," the old prospector replied, "the way it is."

Oddly, anyone floating down the Nahanni reaches the three canyons of the lower river in reverse order, arriving first at Third Canyon. Shafts of sunlight pierced a brooding sky as we drifted into the canyon, a 22-mile-long stretch of sloping hills. We camped that night in the shadow of Third Canyon's most famous landmark, The Gate. Here the river squeezes between a pair of 700-foot cliffs. Attached like a thumb to one wall is Pulpit Rock, a water-sculpted monolith that guards the narrow entrance to the rest of Third Canyon.

We took a break from river routine the next day and lolled in the bright sunshine. Dirk and Derek baked bread; Bo fished for arctic grayling and trout in a nearby creek; and I wandered through a thicket, snacking on raspberries. Later, Derek and I followed Bo's fishing stream for a few miles, and came upon a stunning, nameless gorge. The beautiful little creek raced along a sinuous bed and tumbled down terraced waterfalls. Back at camp we found Dirk cooking rocks! We steamed in a homemade sauna that afternoon, and took lightning-fast dips in the frigid creek.

Darkness comes late to the Nahanni in summer, so that evening we were back on the river. Relaxed by the sauna, we floated along, tired and lazy. It was very quiet. Only the rustle of silt washing against the bottom of

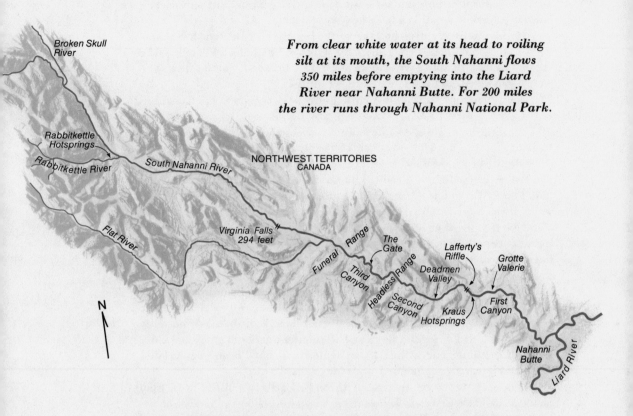

Broken Skull River

Rabbitkettle Hotsprings

Rabbitkettle River

South Nahanni River

NORTHWEST TERRITORIES
CANADA

Flat River

Virginia Falls 294 feet

Funeral Range

The Gate

Lafferty's Riffle

Grotte Valerie

Third Canyon

Headless Range

Deadmen Valley

Second Canyon

First Canyon

Kraus Hotsprings

Nahanni Butte

Liard River

N

the canoe—a sound like bacon sizzling—kept me awake. Forests of spruce and poplar slipped by, and patches of aspen, birch, and jack pine. Mountains loomed in the distance, the Headless Range, site of Second Canyon.

It was nearly dark, and a radiant moon was flashing between the peaks when we pitched camp on a long, narrow island sprinkled with spruce and jack pine. A roaring fire vanquished the chill, and I slept that night blanketed by a multitude of glittering stars.

From atop a mountain the next day, I saw Second Canyon sprawl before me: a platoon of 5,000-foot peaks bisected by a single, meandering brown thread. I thought it ironic that the mountains so dwarfed the river that had carved the gorges. "Canada's finest river canyons," Dr. Derek C. Ford, Chairman of the Department of Geography at McMaster University in Hamilton, Ontario, had called them.

A charming and articulate Englishman, and a veteran of two Nahanni expeditions, Dr. Ford had talked with Bo and me before our trip. "The South Nahanni River channel existed long before the mountain ranges that the canyons bisect," he said. "As the ranges were uplifted, the river maintained its course like a giant knife. The mountains were sliced like slabs of butter being slowly and steadily pressed upward against a blade."

Knives of another sort played a role in Deadmen Valley, the broad, forest-covered lowland where the ominous legend of the Nahanni was born. Charlie McLeod christened the valley in 1908, on the day he discovered the skeletons of his prospector brothers, Frank and Willie, who had been missing for three years. The prospectors had been robbed of their

171

gold, and their heads appeared to have been lopped off. An investigation by the Royal Canadian Mounted Police concluded that the men had been ill-prepared for the harsh weather and had probably starved. Scavengers could have been responsible for scattering their remains. But sensational-ist writers picked up and embellished the story.

Patterson finally dispelled the myth. He traveled the South Nahanni without incident in the 1920's and later wrote of "high pastures of the wild sheep," and of "foaming six-foot waves." He concluded that the area's sin-ister reputation was "much ado over very little."

An S-turn and a gravel-bar riffle welcome canoeists to the Nahanni's finest gorge—First Canyon. Limestone and dolomite walls rise 3,000 feet above talus slopes. Mosses and stunted firs grow on the ledges. When twilight falls, the tan and orange cliff faces slowly turn blue and gray.

Many caves have been discovered in First Canyon's cliffs. In the early morning we joined Park Warden Lou Comin and his assistant Chris Ham-mond on a hike to one of the largest and most complex, Grotte Valerie. It was a three-hour climb to the main entrance, 1,500 feet above the river. "The caves were discovered in 1971," Lou said when we reached the gap-ing west portal. "Grotte Valerie has some 6,300 feet of passageways. We haven't opened them to the public yet; we want to finish some research projects inside before we let too many people in."

We wandered through the cold galleries, dodging beautifully formed stalagmites and stalactites. At the base of an icy, sloping ledge inside the cave we saw the skeletons of more than a hundred Dall sheep. Apparently they had skidded down the slope and were then unable to climb back up. Some of the bones, according to carbon-dating, have been preserved in this frigid permafrost chamber for as long as 2,000 years.

Back on the river we rushed through Lafferty's Riffle, our final obstacle in First Canyon, for awaiting us just outside the mouth of the gorge were the Kraus Hotsprings. They are another vestige of the energy that uplifted, and is continuing to uplift, the Nahanni region. The rancid smell of sulfur did not deter us from luxuriating in the bubbly, 95-degree water, as we had our first warm bath in weeks.

Two days later, the trip finished, we were on a plane headed for Ed-monton. It was early evening and outside my window hung a brilliant full moon, a giant pearl perched above a layer of fluffy, silver-gray clouds. Bo tapped me on the shoulder and pointed to the window on the other side. There the setting sun blazed gloriously. I savored the sight because it was something special—like Nahanni.

Train No. 1 of the Algoma Central Railway eased out of the Sault Ste. Marie depot and chugged northward toward the lakes, mountains, and forests of northern Ontario. Bo Rader and I sat on the dusty floor of the baggage car, munching sour apples and watching through a broad, open door as a panorama of trees and lakes whizzed past.

Agawa Canyon

Most people who see Agawa Canyon do so from the train. Every day, from late May to early October, it comes up from Sault Ste. Marie and runs for ten miles through the canyon. It makes a two-hour stop in Agawa Can-yon Park—time enough for a picnic. During the winter the train makes a run to the park every weekend.

The Agawa River meanders through the canyon on its way to Lake Su-perior, and Bo and I got off the train and hiked the tracks for a few miles alongside the placid waters to the train station. We were met there by Ed Foote, affable Agawa Park Attendant. "Most of the people who come to the park are from the U. S.," Ed told us over dinner that night. "And the numbers have increased dramatically since the railroad began seriously promoting the park. When I came to work here in 1969 we had 30,000 visitors. In 1978 we had more than 100,000."

One damp evening, after a day of steady rain, we borrowed Ed's canoe and paddled downstream toward Bridal Veil Falls. The river, dark as cola from beds of leaves on the bottom, barely moved, and we glided silently along as scores of barn swallows darted above. Here the canyon's east wall, a gray and pink mass of granite, hugs the river and rises 800 feet straight up. To the west, forests of birch, pine, maple, and spruce grow on the canyon floor and cover most of the gentler western wall.

At the base of the falls, a tiered, lake-fed torrent, we fished for trout, but without much luck. Ojibwa Indians once fished here, presumably with better results, but of them only their word for the canyon—*agawa*, the sheltered place—remains.

Fraser Canyon

In 1808 Simon Fraser wrote, "We had to pass where no human being should venture." At the time, Fraser was an agent of the North West Company—the Montreal fur traders. He and a party of 23 men were exploring western Canada when they came upon the canyon and river that now bear Fraser's name. They had to abandon their canoes and grope their way along granite cliffs above the seething river.

Half a century later, gold was discovered in Fraser Canyon—about a hundred miles inland from Vancouver. Thousands of miners poured into the canyon, and a larger strike farther north led to the construction of the 380-mile-long Cariboo Wagon Road in 1864. Today parts of the old road can be seen from the Trans-Canada Highway as it threads Fraser Canyon.

To see the places "where no human being should venture," however, it's best to do as Bo and I did: Take a raft trip. On a scorching midsummer morning, we joined a group of 26 people in two motorized rafts that looked like giant bananas lashed together. As we drifted, one of the guides, Jim Rankin of Whitewater Adventures, told me that the length of Fraser Canyon is something between 27 and 55 miles, depending on how you measure it. "Everyone has a different opinion on where the canyon begins and ends," he said, "because it's not consistently well defined."

To me, the most spectacular section begins near the logging towns of Boston Bar and North Bend. The towns are on opposite sides of the river and are connected by an aerial auto ferry. As we passed beneath its cables, I could see the timbered slopes of the Coast Mountains begin to squeeze the river, and jagged rocks jut up from its turbulent surface. Glaciers moving through the mountains carved a valley along an ancient fault here, and the Fraser River has since dug a canyon in its floor: a U-shape with a 100-foot-deep nick in its bottom. Through this little channel, drainage from 89,000 square miles of British Columbia, nearly a quarter of the province, rushes toward the sea.

At Hells Gate, seven miles downstream from North Bend, we rushed along with it; the current here flows at 25 feet a second between walls of granite just 120 feet apart. An engineering mishap in 1914 further constricted the gate. Blasting sent tons of rocks sliding into the river during the construction of a railroad line.

Tourists in another aerial tram bobbing overhead waved as our rafts entered the surging, leaping waters of the rapids. Immediately, an enormous wave crashed aboard, drenching everyone. We skirted a swirling, hungry whirlpool, and our rafts bucked like broncos through a series of

173

choppy waves before lurching into saner waters again. It was a thrilling, memorable ride, over all too soon.

Later that day we passed small groups of Indians fishing along the banks. The Salish tribes have been netting and drying salmon for winter food here since long before Simon Fraser's time. Now, with Fraser Canyon tamed by railroads, outfitters, and highways, it was comforting to see that at least one element of the canyon's past survives.

Wood Canyon

Boatman Virgil Napier revved his engines and shouted above the roar: "Ready?" Bo and I nodded and huddled deeper into our jackets, for a brisk breeze was riffling the surface of Alaska's Copper River. With the throttle wide open, we thundered out into the swift current of the Copper. Beside us rose the greenstone walls of Wood Canyon, a three-mile-long chasm in the craggy Chugach Mountains near the old railroad town of Chitina, 62 miles northeast of Valdez.

It was a crisp autumn evening when we saw Wood Canyon for the first time. Virgil cut his engines to a whisper, and we idled through the soft twilight. At a narrowing in the canyon, I thought of U. S. Army Lt. Henry T. Allen and his 1885 journey up the Copper. "In places the river does not exceed 40 yards in width and so zigzag is the cañon that in several of the chambers it is difficult to tell the course of the river." He thought the canyon "one of the most picturesque pieces of landscape I have ever seen," and named it for his commanding officer, Col. H. Clay Wood. At the head of Wood Canyon, Lieutenant Allen stopped at an Indian village and was welcomed by Chief Nikolai. The Indians fired a salute to the soldiers, and Allen, examining their weapons, found that they were using bullets of silver-copper alloy. The metals came from mines nearby.

Fourteen years later, the Indians revealed the whereabouts of the mine to a group of Klondike prospectors, and the rush was on. By 1911, an Alaskan syndicate had laid 200 miles of railroad track across rivers and mountains, linking the mines at Kennecott to the coast at Cordova. By 1938, when the mines closed, nearly 600,000 tons of copper and 3,000 tons of its by-product, silver, had been mined at Kennecott.

We spent the night at Virgil's snug camp a few miles beyond the southern entrance to the gorge, and in the morning we hiked back through the canyon on the Copper River Highway, a one-lane dirt road that replaced the abandoned railway. Even the road is closed to vehicles now—the result of rockslides and bridge washouts in 1975—but it remains the only land route into the canyon.

In places the dusty road hung precariously close to the edge of the cliff, and we peered straight down at the seething Copper River. From a moss-covered ledge we viewed the canyon's most elegant scene, the waterfall of Tenas Creek, a gleaming thread that plunges more than 100 feet down the opposite wall. And in the shadows of a waning Alaskan summer sun we came across the remnants of an old railway—dilapidated trestles, wooden monuments to other generations. As we left Wood Canyon, I realized that despite my lack of wilderness experience I had found nothing to fear in the canyons: neither in the Nahanni, nor Agawa, nor Fraser. Only a newfound exhilaration in the legendary northern wilds.

Dying sun dusts snowcapped peaks of the Chugach Mountains overlooking the Copper River's Wood Canyon in southern Alaska. Indians once mined silver and copper here; now weekend fishermen line its riverbanks in summer, dip-netting for salmon.

Agawa Canyon

In the dense forest of Lake Superior's eastern shore, an Algoma Central train hugs a bank of the Agawa River in timeworn Agawa Canyon. The line serves miners and loggers working out of Sault Ste. Marie in a 1,300-square-mile wilderness area owned by the railroad. Throughout the year, special tour trains bring thousands of visitors into Agawa Canyon for brief stops. Visual delights such as Bridal Veil Falls (right), and inky cap mushrooms, await canyon guests.

177

Fraser Canyon

Granite precipices pinch the churning waters of
the Fraser River in British Columbia's Fraser Canyon.
Salish Indians have fished these waters for centuries,
trading smoked salmon to tribes farther inland. After
netting (center, left) and filleting (left), the fish hang
on wooden racks to dry. "Drying normally
takes three weeks," says Mrs. Bertha Peters, who
has repeated the process every summer for 33 years.

Sharp ridges, carved from ash and lava by rushing streams, rise half a mile above the floor of Waimea Canyon on the Hawaiian Island of Kauai.

Hawaii:

Waimea Canyon

Photographed by National Geographic Photographer George F. Mobley

Mohihi Stream plunges hundreds of feet to the floor of a side canyon in Waimea.
Tropical flowers, many brought from other countries, flourish in the
canyons. At upper right, Mauritius hemp, a relative of the century plant introduced
from South America, develops a little plant at the base of each pale-green
flower; when the blossoms drop off, the plantlets remain and continue to grow (upper left).
At lower left, a coiled tendril hangs above the bloom of a banana passion
fruit vine, another native of South America. A hybrid member of the iris family
(lower right), first cultivated in Europe, lifts its three-inch-wide
flower on a plateau above Waimea Canyon. Such imported species crowd out Hawaii's
native flora; 40 percent or more of the islands' original plants face extinction.

Picking his way across a brook, a youngster explores the edge of Waimea Canyon. Such streams have cut a network of ravines into the canyon's east rim (on the right). They originate in the Alakai Swamp, where rainfall averages up to 450 inches a year. Here on the leeward side of the island, just a few miles from the swamp, the land remains much drier. The mouth of Waimea Canyon receives only about 25 inches. Prickly pear cactus (left), introduced from the mainland about 1800, spread throughout these drier portions of the islands. A highway on the west rim of the canyon provides access for visitors, such as the family opposite. More than a million people a year view Waimea from lookouts, but only a few make the arduous two-and-a-half-mile descent to the canyon floor.

On a slope made treacherous by loose fragments of volcanic rock, author Bob Morrison hikes into Waimea Canyon. "We tried to follow a little-known hunters' track," he recalls, "but it soon disappeared, and we had to make our own trail for most of the way." That evening he and photographer George Mobley set up their tents (above) on terraces built by early Hawaiians for growing taro, a potato-like staple. "The terraces covered every bit of ground along the little stream," Bob says, "ending only at the foot of a cliff." To irrigate their crops, which required great quantities of water, the Hawaiians constructed dams and extensive networks of ditches. At least one ditch remains in use today, incorporated into a system irrigating fields of sugarcane.

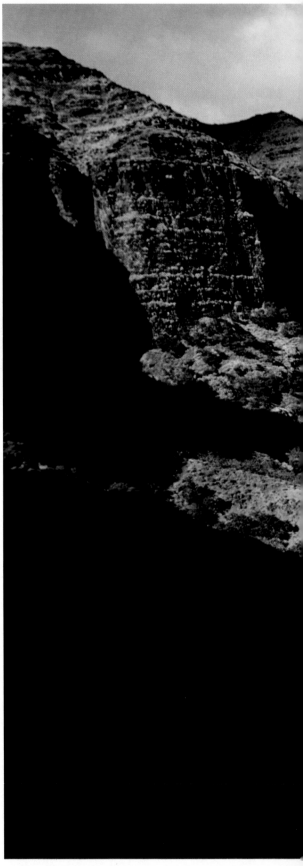

Boulder-strewn bed of Koaie Stream winds toward the Waimea River. In Hawaiian, *waimea* means "reddish water," an apt description of the silt-laden river. On a cliff face 800 feet above the canyon floor, geophysicists Scott Bogue, at left, and Mark Linker study the magnetism of the earth. Using an instrument that combines compass and level, the two record the positions of core samples of lava that cooled and solidified millions of years ago. Scott explained: "Analysis of the faint magnetization which these lavas acquired soon after eruption will tell us something about the movements of the earth's magnetic poles."

By
H. Robert
Morrison

I decided to jump. From the ledge where I stood, halfway down the side of the canyon, there was only a short drop to what seemed to be solid footing. A steep slope, studded with boulders and prickly pears, stretched for a thousand feet below. My boots hit the loose volcanic soil and skidded out from under me. Head over heels I tumbled down the slope, dangerously out of control.

With National Geographic photographer George Mobley, I was exploring Waimea Canyon on the Hawaiian Island of Kauai—the first of four canyons we would visit in Hawaii. We had chosen a little-known route described by a park official as a "hunters' track." It was not marked on any map. The day before, George and I had picked out the route through binoculars, but we had not seen the ledge.

When we reached it, George removed his backpack and climbed down a cleft between two rocks. He lifted down his pack, then took mine. Without the extra 50 pounds, I felt the short jump would be easy. Now I knew how wrong I had been.

Desperately, I dug my hands into the earth. Dirt ripped at my fingernails. Sharp rocks scraped my palms. But my rolling slowed. Again I clawed at the ground. This time I stopped and lay trembling.

A quick check: no surges of pain, no broken bones, no sprains. Even my glasses remained intact. I shouted to let George know I was all right, slowly stood up, and trudged back to where he waited. I sat down with my back to the ledge and took a long drink of water from his canteen. As we rested, my panic gradually ebbed, and once again I could begin to appreciate the beauty of Waimea Canyon.

The predominant color of this half-mile-deep gash in the earth is rusty red, but layers of dark gray and yellow are visible here and there on the nearly vertical walls. Patches of green show where plants keep a precarious hold in the hot soil. At the bottom of the gorge, a strip of vegetation marks the course of the Waimea River. To my right I could just make out the silvery thread of a waterfall at the head of the canyon. As I watched, clouds drifted over, and the changing colors in the shifting sunlight and shadow made the rocks themselves seem almost alive.

But the sun was dipping closer to the rim, and we wanted to reach the stream before dark. We shouldered our packs, and I followed George as he switchbacked down the slope. We had almost reached the thickets at the bottom when we spotted a goat on the opposite wall. It came into view around an outcrop of rock, walking an impossibly narrow ledge high above us. Another followed, then another. We counted 14 of the animals, defying gravity as they walked single file along the face of the cliff.

We would see many more goats during our hike. Except for the soaring white-tailed tropic birds, which delight visitors at the overlooks on the rim, the goats are the most visible wildlife in Waimea. They are the wild descendants of domestic goats. Because their browsing is destructive to plants and encourages erosion, goat hunting is permitted in Waimea on weekends during August and September.

I would see how destructive they could be during a later trip to the canyon, a brief but exciting helicopter ride. Pilot Jack Harter knew the canyon's geography intimately. For half an hour we viewed Waimea from a new perspective, sweeping down side canyons that ended in amphitheaters

Solitary ohia-lehua tree clings to a fluted wall of Waimea Canyon. Early Hawaiians carved images from the wood of the ohia-lehua, the most common of the islands' native trees.

191

set with waterfalls, each more spectacular and breathtaking than the last. Several times we landed on flat spots, some of them no bigger than tennis courts, atop spires and ridges with sides dropping sheer for half a mile to the canyon floor. On one ridge Jack pointed to a fence enclosing an area a few yards square. "It was built several years ago to see what would happen if the goats were kept out," he told us. Outside the fence there was nothing but crumbly rock; within grew a carpet of grass, and a small tree had taken root. "I believe Waimea would be a much greener canyon if only the goats could be controlled," he said. Considering the contrast between the vegetation inside the fence and the bare rock outside, I had to agree.

Now, lengthening shadows urged us onward, and we soon reached the canyon floor. It was time to look for a campsite. I was struggling through brambles and passion fruit vines when I heard a whoop from George. He was smiling broadly as he emerged on the opposite bank.

"Just wait till you see what I've found," he announced with a grin. I followed him a short distance downstream, clambered over some boulders, and found myself standing on a remarkably flat area.

"Look there, and there," George said, pointing.

Walls of black, uncut stone held level plateaus, terrace upon terrace, reaching upward from the stream as far as we could see. "Welcome to the ideal campsite, courtesy of the ancient Hawaiians."

I had read of such sites in side canyons farther downstream, but I hadn't expected to find evidence of early settlement at the very head of the main canyon. Why, I wondered, would anyone settle here, with fertile, easily tilled land along the coast just a dozen miles away?

I asked the question later when I talked with Dr. Yosihiko H. Sinoto, Chairman of the Department of Anthropology at the Bernice P. Bishop Museum in Honolulu. "Most likely because of population pressures," he told me. "From similar sites excavated in other valleys, we know the earliest settlements were usually built at the mouths of the valleys. As these settlements grew, they expanded inland. To increase the amount of farmland, the Hawaiians built terraces on the valley slopes.

"We can't say for certain how old the Waimea sites are," he continued, "because we haven't made extensive excavations there. But I would estimate that settlement began in coastal Waimea in the 14th century, and expanded inland during the next 200 years." I knew that most of the remote valleys and canyons were deserted, and I asked Dr. Sinoto why.

"The main causes were Christianity and the introduction of a money economy, resulting in trade and urbanization," he said. "To trade with the European sailing ships, the Hawaiians had to go to the coastal harbors. This could mean a journey of several days each way. Christianity was a factor because it emphasized weekly church attendance. And that, again, was difficult for people living in the valleys."

That first night I spent in Waimea Canyon seemed far too short. I felt I had just eased my aching body into my sleeping bag when a scream jerked me awake. I lay still for a moment, blinking in the early-morning sunlight that filtered into the tent, trying to collect my thoughts. Had I been dreaming?

The scream sounded again, a rasping shriek, and I poked my head out of the tent and peered around. Nothing. Pulling on my trousers and boots, I cautiously stepped toward the stream. A sudden whirring of wings startled me, and I glimpsed a brownish bird about the size of a large chicken— apparently as unnerved as I. It was an Erckel's francolin, introduced

Mohihi
Stream
Koaie Stream
Alakai
Swamp
Koaie Canyon
Trail

*Born of fire
and water—
volcanism and
erosion—
Waimea
Canyon slices
southward
near the west
coast of the
Hawaiian
Island of
Kauai.*

as a game bird and now common on Kauai. I was later amused to find that a book on Hawaii's birds described the male's call as a "cackle."

After breakfast we continued down the floor of the canyon. No trails extended this far upstream, so we followed the river, our muscles protesting as we made long steps from one boulder to another. At intervals, steep spurs of the crumbly volcanic rock rose in our path, and we crossed and recrossed the river on stepping-stones to get around them.

The burning sun lifted higher in the sky. A bandanna sweatband I had tied around my head was soon soaked through. Now and then a faint breeze stirred the air, and I loosened the straps of my pack to let it play across the back of my sopping shirt. Kukui trees grew everywhere, their silvery foliage glinting in the bright sun. The kukui, or candlenut, is the state tree of Hawaii. It was brought to the islands by Polynesian settlers. They strung the kernels, which have a high oil content, on the ribs of coconut or palm leaves and burned them as candles. Walking beneath the trees, we stepped carefully. Fallen nuts and husks lay in a thick litter, and on the smooth rock were as slippery as marbles on a sidewalk.

The going was faster when we reached the Waimea Canyon Trail the next day, and we stopped to rest at a dam across the river. Beside a flea-infested vacant hut, its iron roof rusting into shades that matched the canyon wall, grew an orange tree. We savored the sweet juice as we sat beside the dam. Only a trickle of water ran down the spillway; most was diverted into a tunnel cut into the rock. The water would flow to a power plant and from there to fields of sugarcane, the island's most important commercial crop. Vast amounts of water are needed in growing and processing the cane—altogether, some 250 gallons, or one ton of water, to produce one pound of refined sugar.

Below the dam we found the Koaie Canyon Trail leading up a side valley. To the muted thunder of the Koaie Stream tumbling along its rocky bed, we hiked the shady trail. When we rounded a spur, a break in the trees revealed a vista I would not have expected to see in Hawaii. Above the green border of trees on the narrow valley floor, sheer slopes of red rock thrust more than a thousand feet toward the passing clouds. Below us ran the sparkling rivulet, its course punctuated by waterfalls.

As we proceeded, the sharp scent of ginger now and then filled the air; we often smelled the plants before we could see them. Kona coffee trees grew in thickets, some of them well over 25 feet tall.

At one point George stopped suddenly, bent down beside the trail, and picked up a ripe avocado. I glanced up and saw that the leaves of several tall trees matched those of the tiny avocado plant growing in my living room. George was already foraging and in a few seconds sat down with two more ripe avocados. He looked so eager and happy I hesitated to tell him that they aren't among my favorite foods. At the first bite, a dreamy expression spread across his face. "Just lying around," he mused. "A costly luxury back home, but here just lying on the ground rotting away."

My turn to savor the wild foods—and George's turn to joke about dreamy expressions—came that evening when we camped beside an enormous papaya tree.

The next morning, our last in Waimea Canyon, we dawdled over breakfast and lunch; we knew the hike out would be hot enough without the midday sun. The Kukui Trail climbs 2,000 feet from the floor to the rim in just two and a half miles. For most of the way, there is little shade.

At first the climb was easy. Our legs, conditioned by three days of backpacking, were in good shape. The food we had eaten had lightened our packs. But the trail led relentlessly up. Our pace slowed. I began to feel as though I were carrying three heavy bags of groceries up a sweltering, endless staircase. The steel trail markers set every quarter mile seemed to grow farther and farther apart. Upward, still upward, we toiled, stopping to rest whenever a turn in the trail or an outcropping of rock cast a sliver of shade.

As we neared the rim, we began to see *iliau*, shrubs which grow on the drier uplands of Kauai and nowhere else on earth. Large, ball-like clusters of long, tapered leaves top thin stalks as much as ten feet tall.

We reached the canyon rim late that afternoon. I was hot. I was thirsty. I was hungry. I was exhausted. I hated my backpack, and couldn't wait for the luxury of a hot, lingering bath. One last glance at the canyon, I thought, and I would head for the car.

But that last glance turned into a long look. Lengthening shadows were stealing up the gorge. Wispy patches of cloud threw fingers of shade along the walls. Two tropic birds soared far below me, their long white tail streamers trailing gracefully. A shaft of sunlight played like a spotlight over a cluster of kukui trees. The beauty of the scene held me fast.

To learn how Waimea Canyon was formed, I spoke with Dr. Doak Cox, a geologist who heads the Environmental Center at the University of Hawaii. "The geological history of Waimea is very complex," he said, "for the canyon you see today represents several stages of development." He paused for a moment to fill his pipe with fragrant London Dock tobacco, then went on to speak to me of periods of volcanism and of volcanoes that collapsed; of cycles of erosion; of calderas and "grabens"—depressions in the earth's crust bounded by faults; and of a time of alluvial accumulation, all of which went into the creation of Waimea Canyon.

"As the prevailing winds sweep across the island from the northeast," Dr. Cox continued, "they drop their moisture mainly on the windward side and on the Alakai Swamp, creating one of the wettest areas on earth. At the high point of the swamp—Mount Waialeale, only ten miles east of Waimea Canyon—the average rainfall is more than 450 inches a year. The rainfall averages only about 40 inches at the head of Waimea Canyon, and about half that at its mouth. The valleys on the windward side are quite different: They're wet."

D r. Cox was right. The differences were immediately obvious as George and I drove a rented jeep into Wainiha Valley, on the windward side of Kauai. A sugar company controls most of the valley, and uses the power of the Wainiha River to produce electricity. The steep, fluted walls of the valley were lush with greenery, broken only where the thin threads of waterfalls dropped hundreds of feet toward the river winding below. Ferns and mosses everywhere gave silent testimony to the abundance of rainfall. Thick clusters of ginger flowers— both yellow and white varieties—mingled their sharp scents with the dull smell of wet, decaying vegetation. Stalks of Philippine ground orchids stood beside the road, each pink flower a study in miniature perfection.

Parking the jeep beside the dam at the end of the road, we hoisted our packs and set out to find a trail up the river. We had seen on a map a gauging station a mile or so upstream, and reasoned that a trail would lead to it. We finally found the trail after half an hour's search—more accurately, we found the faint trace of it. As we bushwhacked our way up the valley, the track proved impossible to follow. We would find it for a few yards, then lose it again. When we tried to hike along the rocks at the river's edge, we

Wainiha Valley

found that the water level was too high. Also, rain had begun to fall, and the mossy rocks were dangerously slick.

This was a tropical rain forest, the first I had ever seen. Ferns grew shoulder high. Giant tree ferns towered above them, with ten-foot fronds arching from the trunks, and fist-size fiddleheads reaching up from the centers. Above all towered the trees, a green canopy that blocked the sky but let in the rain.

For hours we struggled along, stumbling over rocks and fallen logs concealed in the lush undergrowth. By the time the afternoon light began to fade, we had gone less than a mile. There in the rain, with mosquitoes zeroing in on us whenever we stopped for a breath, we decided to head back to the jeep.

Our defeat was only temporary, thanks to Jack Harter and his helicopter. As we walked toward the chopper on its landing pad at the Kauai Surf Hotel, we mentioned our difficulties in following the trail.

Jack laughed. "Nobody uses that trail anymore. When the gauge needs checking, I fly the men in."

Spatters of rain flecking the bubble of the helicopter at Wainiha underscored the differences between this valley and Waimea Canyon. We circled slowly at the head of Wainiha, a vast amphitheater with a majestic waterfall plunging hundreds of feet to a pool enclosed by the ever-present green of the valley. Even the rocks at its edge were velvety with moss.

For as long as we could, we lingered there, savoring the play of cloud against curtained wall, and the black silhouettes of ohia-lehua trees against a soft gray background. Finally, Jack lifted us above the valley walls, and we headed back.

Akaka Falls State Park

To reach our next destination, George and I took a plane to the "Big Island"—Hawaii. There we talked with Seiso Kamimura, Park Superintendent for the Hawaii District, in his office at the State Building. Over coffee, he described the attractions of Akaka Falls State Park.

"The falls themselves are interesting," he began. "After all, one of them is 420 feet high, and with the extra flow from the rain we've been having, it should be more spectacular than usual. However, the falls are not the only reason that more than 800,000 people visited the park last year.

"In the first place, the park is easily accessible. You don't have to go backpacking or bushwhacking. The trail is paved, and it's less than half a mile long. In the second place, plants from all over the world thrive in the junglelike atmosphere of the park. There's hardly a square foot that doesn't have some kind of flower."

The park at Akaka Falls was all that Mr. Kamimura had promised. A gentle trail wound through a tropical garden, with flowers ranging from modest impatiens to incandescent displays of red ginger, thrusting like fiery ladders amid the green swords of their leaves. I was especially fascinated by a grove of enormous bamboo; some of the stalks were more than six inches thick. I felt as though I were walking through a Japanese print.

First I passed Kahuna Falls, which drops from a side canyon. Spectacular in its own right, Kahuna is overshadowed by the majesty of Akaka Falls farther along.

The roar of falling water grew louder as I strolled down the gently sloping trail. Sure enough, the rainfall had swollen the stream, and its water surged from the cliff top. As it fell toward the roiling pool, gusts of wind lifted wisps of spray and sent them flying. When the clouds parted for a moment, a pastel rainbow glowed in the mist, then softly faded.

As I leaned against a railing, I overheard a youngster beside me remark to his mother, "Boy, wouldn't it be neat to ride down that on an inner tube—but you'd need a parachute, too!"

The gathering clouds had begun to fulfill their promise; a light rain was falling as I walked back. I had almost reached the parking lot when I happened to look upward. On one of the spreading branches of a giant albizzia tree grew a cluster of three hybrid cattleya orchids. Here and there a crystal raindrop trembled on a petal's edge. The showy purple flowers contrasted vividly with their background of dark foliage and gloomy sky, my last memory of Akaka Falls State Park.

Waimanu Valley

Luckily, there were no permanent settlements in Waimanu Valley in 1946 when a giant tidal wave swept in. For centuries before that, this valley on Hawaii was occupied. The tsunami destroyed modern residences and ancient terraces alike, though a few archaeological sites survived the wave and can still be seen.

In 1873 Isabella Bird, an adventurous Victorian traveling for her health, explored the islands on foot, by carriage, in a sedan chair, and by horseback. She managed to reach the falls of Waimanu, with her horse "now and then hanging for a second by his fore feet." She described the valley as "one perfect, rapturous, intoxicating, supreme vision of beauty."

The valley has been little disturbed in recent years. The isolation is one reason that Waimanu is being made a National Estuarine Sanctuary. Only a few backpackers and hunters visit the valley. Hunting of the wild pigs is encouraged, because their rooting threatens plants and archaeological sites, and hastens erosion.

One afternoon George and I pitched camp in a grove of ironwood trees at the edge of the cobbled beach at Waimanu Bay. Picking our way along the charcoal-gray boulders that afternoon, we saw a man hunched at the water's edge, peering intently at incoming waves. Suddenly he rose and, in a single fluid motion, flung a net into the water. When he stepped back to the beach, a fish was flopping in the mesh.

"Do you get a fish every time you throw?" I asked.

"No," he replied with a smile. "Usually more!"

We laughed and introduced ourselves. His name was Jacob K. Batalona, and he was serving as a guide to a party of mainlanders here on a sight-seeing trip.

"I come here often," he told us. "My mother-in-law was born in Waimanu Valley." I knew that a number of local residents had opposed the estuarine sanctuary. Many have an emotional bond with Waimanu Valley because their ancestors lived there. Most also felt they were not receiving fair prices for their land, and they resented the new regulations governing land and water use, and restrictions on camping. I asked Jacob how he felt.

"At first I was against it," he said. "I wanted Waimanu to stay as it had always been, so my children could enjoy the unspoiled valley as I do. Then I heard rumors of plans to open a luxury resort here. I came to realize that, if the valley were to be preserved, it needed protection." He gazed for a moment down the long green expanse. "Perhaps the sanctuary will be a good thing—good for my children, and for generations to come."

Silvery waterfalls thread the walls of Waimanu Valley, a canyon on the windward side of the Island of Hawaii. Craggy, 1,200-foot cliffs, clothed in lush vegetation, border the valley's table-flat floor.

Wainiha Valley

Tangled rain forest mantles cloud-wreathed
Wainiha Valley on the windward side of Kauai. At
left, a native hibiscus tree blooms in a thicket on the
damp valley floor, and a damselfly alights on a
pebble. White ginger blossoms atop stalks five feet tall
emit heady perfume; pink Philippine orchids,
two inches across, have no fragrance.
Following pages: Fist-size fiddlehead, the frond of
a tree fern, unfurls in Wainiha Valley. Eventually
it will grow as much as six feet long.

Akaka Falls State Park

Akaka Falls plunges 420 feet into its gorge, where other wispy cascades tumble. Exotic flowers transplanted from around the world bloom here: Hybrid cattleya orchids (below, left) grow on a tree branch; heliconia displays a ladder of showy bracts. Red ginger (bottom, left) came from other Pacific islands, and torch ginger from farther east, probably Indonesia.

Notes on Authors

TOR EIGELAND, a native of Norway, studied at McGill University in Montreal, at Mexico City College, and at the University of Miami (Florida) before becoming a free-lance photojournalist. His work has taken him all over the world. *Isles of the Caribbean*, to be published by the Society in February 1980, will include his coverage of Trinidad and Tobago. RALPH GRAY has been a National Geographic editor and writer for 36 years. While Chief of School Service, he edited the *School Bulletin* and wrote the Old Explorer column that appeared in it. In 1975, when *World* magazine replaced the *Bulletin*, he continued as editor of the new children's monthly. WILLIAM R. GRAY, a graduate of Bucknell University, joined the Society's staff in 1968. He has written *The Pacific Crest Trail*, chapters for a number of other Special Publications and *Camping Adventure*, a book for children. His book about Captain James Cook will be published by the Society in February 1981. H. ROBERT MORRISON, a writer and editor, has contributed chapters for a group of Special Publications ranging from *The Ocean Realm* to *Mysteries of the Ancient World*. He has been a staff member since 1964. THOMAS O'NEILL worked as a newspaper reporter in Wisconsin and as a writer in Washington, D. C., before joining National Geographic in 1976. In August 1980 the Society will publish *Back Roads America*, a book for which Tom is currently doing fieldwork. MICHAEL W. ROBBINS grew up in Ohio, received his B.A. from Colgate, his M.A. from Johns Hopkins, and his Ph.D. from George Washington University. He wrote a chapter on mountain men for the Special Publication *Into the Wilderness*. GEORGE JOSEPH TANBER has worked as a writer in Washington, D. C., and in Beirut, Lebanon. He holds a master's degree in journalism from Ohio University. His chapter in *America's Majestic Canyons* was his first Geographic assignment. EDWARD O. WELLES, JR., is a 1972 graduate of the University of North Carolina. He contributed a chapter on William Bartram for *Into the Wilderness*. His work has also appeared in the *Washington Post*.

Acknowledgments and Additional Reading

The Special Publications Division is grateful to the individuals, organizations, and agencies named or quoted in the text and to those cited here for their generous assistance: John P. Bluemle, William Bock, Bill Bonnichsen, Duncan Burchard, Robert A. Bye, Jr., Jack Carlson, Nick Carter, Frederick Cassidy, Catharine Castro, Robert Chapman, Dave Conrad, Dan Culver, John Fay, Donald Fisher, Jacques Fleury, Roger Giddings, Roselle Girard, Stanley C. Grant, Eliette Grenier, William C. Griggs, Peter Harker, Douglass Henderson, Richard Keefer, Charles Lamoreaux, Ed MacKevett, Tim Manns, Ross A. Maxwell, David May, Maynard M. Miller, Jack Musgrave, Campbell W. Pennington, Richard Preston, William B. Purdom, Alan Rayburn, Eugene Roseboom, Gregory Sassaman, Robert H. Schmidt, Jr., Henry W. Setzer, Gene Smalley, Jerry Vineyard, Deward Walker, Merle Wells, Peter Wilshusen, and Mary Woods; U. S. Forest Service, National Oceanic and Atmospheric Administration, National Park Service, Parks Canada, Smithsonian Institution, and the U. S. Geological Survey.

R. S. Babcock, et al., *Geology of the Grand Canyon*; Wendell C. Bennett and Robert M. Zingg, *The Tarahumara*; F. S. Dellenbaugh, *A Canyon Voyage*; Francis P. Farquhar, *History of the Sierra Nevada*; Mary Hill, *Geology of the Sierra Nevada*; Elizabeth Hogan, ed., *Rivers of the West*; Charles B. Hunt, *Natural Regions of the United States and Canada*; Bruce Hutchison, *The Fraser*; Joseph C. Ives, *Report Upon the Colorado River of the West*; Michael Jenkinson, *Wild Rivers of North America*; William R. Keefer, *The Geologic Story of Yellowstone National Park*; George E. Knudson, *A Guide to the Upper Iowa River*; Joseph Wood Krutch, *Grand Canyon: Today and All Its Yesterdays*; S. W. Lohman, *The Geologic Story of Canyonlands National Park*; Carl Lumholtz, *Unknown Mexico*; Gordon A. Macdonald and Agatin T. Abbott, *Volcanoes in the Sea: The Geology of Hawaii*; John Muir, *Our National Parks*, *The Yosemite*; Boyd Norton, *Snake Wilderness*; R. M. Patterson, *The Dangerous River*; Campbell W. Pennington, *The Tarahumar of Mexico*; Donald G. Pike, *Anasazi: Ancient People of the Rock*; J. W. Powell, *The Exploration of the Colorado River and Its Canyons*; David A. Rahm, *Reading the Rocks*; Frederick W. Rathjen, *The Texas Panhandle Frontier*; Elfriede Elisabeth Ruppert, *A Historical and Folklore Tour of the Pennsylvania Grand Canyon*; Carl P. Russell, *One Hundred Years in Yosemite*; Harold T. Stearns, *Geology of the State of Hawaii*; G. J. Tucker, *The Story of Hells Canyon*; Dick Turner, *Nahanni*; R. C. Tyler, *The Big Bend*; T. W. Watkins, *The Grand Colorado*. Readers may also wish to consult the *National Geographic Index* for related articles.

DAVID MUENCH

One of America's most majestic gorges, the Grand Canyon sprawls across a corner of Arizona. The Colorado River flows along the canyon floor, 3,000 feet beneath a solitary visitor.

Library of Congress CIP Data

America's Majestic Canyons
Bibliography p. 204   Includes index.
1. Canyons—United States. 2. United States—Description and travel. I. National Geographic Society, Washington, D. C. Special Publications Division.
GB563.A47   917.3'09'44   78-61263
ISBN 0-87044-271-6

Composition for *America's Majestic Canyons* by National Geographic's Photographic Services, Carl M. Shrader, Chief; Lawrence F. Ludwig, Assistant Chief. Printed and bound by Holladay-Tyler Printing Corp., Rockville, Md. Color separations by Colorgraphics, Inc., Forestville, Md.; Graphic South, Charlotte, N.C.; National Bickford Graphics, Inc., Providence, R.I.; Progressive Color Corp., Rockville, Md.; The J. Wm. Reed Co., Alexandria, Va.

Index

Boldface indicates illustrations;
italic refers to picture legends